INSTRUCTOR'S
SOLUTIONS MANUAL

EXPERIENCING
GEOMETRY
ON PLANE AND SPHERE

DAVID W. HENDERSON
Cornell University

with writing input from
Eduarda Moura
and
Justin Collins
Kelly Gaddis
Elizabeth B. Porter
Hal W. Schnee
Avery Solomon

Prentice Hall, Upper Saddle River, NJ 07458

© 1996 by **PRENTICE-HALL, INC.**
Simon & Schuster / A Viacom Company
Upper Saddle River, NJ 07458

10 9 8 7 6 5 4 3 2 1

ISBN 0-13-396813-8

Printed in the United States of America

INSTRUCTOR'S
SOLUTIONS MANUAL

EXPERIENCING
GEOMETRY
ON PLANE AND SPHERE

Dedicated

to all the students who have studied geometry with me

(you have taught me much about geometry)

and to Susan Alida

(you have taught me much about my Self)

Contents

Message to the Instructor

A student is not a vessel to be filled; but rather, a lamp to be lit.
— A. D. Alexandroff

Experiencing Geometry is based on a course for mathematics majors (and others) which I have been teaching at Cornell University since 1974. This book was written specifically for the students in this course, but I have also successfully used many of these problems in a variety of different courses and workshops, including in-service courses for teachers in the USA and South Africa, a course for mathematics majors at the Palestinian Birzeit University, a freshman course at Cornell for students "who are not yet comfortable with mathematics," and even in a course for second graders. Of course, students with different backgrounds and different experiences have responded to the problems in different ways and at differing levels of sophistication. However, since the material is based on experience and imagination, people with differing backgrounds are able to find the material accessible and challenging. In one day-long workshop there were fifty participants ranging from elementary school teachers (with no university level background in mathematics) to a few research mathematicians from the local universities. Each participant was able to grapple with the problems in his/her own way in small groups and then share understandings and insights with the whole group. Again, this text is specifically designed for mathematics majors and prospective teachers — students with less background will probably find the level of discussion in the text too challenging.

It is important that you read the students' text and think about each problem before you read what is written about the problem in this *Instructor's Manual.* This is so you will have some experience of what your students are going through.

In this *Instructor's Manual* there is included a discussion of the variety of the students' responses. For almost every problem there is a verbatim response from a student to the problem. These responses are submitted in writing and then the teaching assistant (T.A.) and I make comments. The students then respond to the comments. This process would continue until both the student and we were satisfied. This variety of student responses is included here to help you hear the ideas imbedded in what your students say and write. They may be awkward and confused in expressing their ideas, but if you listen carefully you will learn much about geometry from them as I have. Even though I have been teaching this course for twenty years, my students still find proofs or ways of looking at the geometry that I have never seen. During 1988-1993, 56 out of the 178 students in the mathematics major's course showed me something new about geometry. This occurs because my teaching style permits and encourages students to explore their own ideas and draw their own conclusions.

Similar experiences will probably happen to you. But you will have to allow yourself to be open to learning from your students. When I started I found this to be very difficult. When a student would show me an argument which I had never seen before, often my immediate (gut-level) reaction was that their argument was wrong, and I felt the urge to tell them the "correct" (i.e. my) answer. Only by careful listening on my part and persistence on the students part was the new way of seeing eventually communicated to me. Often this experience was most difficult (and most rewarding) when it involved a student who was different from me in gender, race, or cultural background. (See in the main text for a further discussion of this.) You may feel as uncomfortable as I was with this approach, but I urge you to try it. The benefits for both you and your students will be well worth the effort. There is a fine boundary between imposing one's own views and guiding a student to see in his/her own way. With time and experience one can begin to recognize the distinctions between the two in different contexts and learn to work around and with in that boundary.

It is important to continually encourage the students to experience their own understandings. You may need to remind them to reread parts of the "Message to the Reader" and "How to Use This Book" and to give individualized guidance where needed. Supplement these sections as you see fit. For this purpose some instructors have found *Proofs and Refutations*, [**L:** Lakatos][†], useful.

I find that it is necessary and important to encourage the students to talk with each other. Sometimes I organize some small groups in class, but mostly I encourage the students to work together outside of class. If someone can't understand something, often times a classmate has another approach to the problem which helps to clarify things for that person. I require that each student write their solutions in their own words, and if they got ideas from others in the class, to acknowledge those contributions the same as we do when we write research papers.

When, as a class, we finished each problem I tried to give some closure to what we had found and some pointers as to where one could go from there. I hope this Instructor's Manual will help you do the same, but the nature of the material and the process will always leave many loose ends and unexplored paths.

The numerous examples of student's work found in this *Instructor's Manual* are used here with the written permission of the students. The students work is included here as they wrote it; the only changes were that the text was copy-edited most of the simple drawings were redrawn by the computer.

May 21, 1995

David W. Henderson

[†]The references in brackets refer to the "Bibliography" in the main text.

Chapter 1

Straightness and Symmetry

You are encouraged to try this problem yourself and to investigate fully all the symmetries of straight lines before beginning this chapter with your students. If you try to do this problem ahead of time, you will hopefully be able to anticipate the places where students will stumble while tackling it. Symmetry will be a crucial element of almost every problem the students explore. Also, construction is fundamental to an exploration of straightness through symmetry. Students must use models as a means of testing their hypotheses, to convince themselves and others. Also, students must construct their ideas about straightness from the ground up, and then continually re-examine those notions and assumptions which, at first, may seem peripheral or incidental. Nothing should be taken for granted. These moments of exploration and re-examination can create opportunities for tremendous student (and instructor!) creativity. The instructor should try to foster this creative spirit as much as possible.

PROBLEM 1. *When Do You Say That a Line is Straight?*

When we ask students this first question, we try to convey it in the following way: "Try to build for yourself a notion of a straight line. For example, think about how you would build a straight line and how you would check if the line you constructed is straight. Consider ways that you might convince someone that the line is, in fact, straight. Look to your experiences."

At first, students will look for examples of the physical world or natural straightness. They are likely to bring up ideas such as using a straight edge, stretching a string, sighting along a line, or, perhaps, following a laser beam. It is necessary then to guide them toward thinking about what is common among all of these "straight phenomena."

As students look for properties of straight lines that distinguish them from non-straight lines, the following statement (which is often taken as a definition) usually arises in class: *"A line is the shortest distance between two points."*

Students may turn to this definition because it is, in fact, a common feature of the straightness expressed in the examples they have examined; perhaps, too, they are compelled to rely on it because this definition is most commonly used in high school mathematics classes. Whatever the case, the following questions may help students take a closer look at this definition:

· Can you always measure all the paths between two points?

· How do you find the shortest path?

· Is the shortest path between two points in fact a straight line?

· Is a straight line between two points always the shortest path?

Students often cite examples of sighting along a line as a way to check for straightness. One example of "sighting" is a technique that is used for laying out fence posts. One can sight along a post on the end to see if any of the others are "out of line." From the view at the end, all the posts should coincide. The notion of sighting along a line comes from a property of light; that is, light always travels along the path that takes the least time. This "least-time" path is straight whenever the light is traveling through a vacuum or a uniform medium such as air at a uniform pressure and temperature. When light crosses into a different medium, such as glass or water, it bends. This is the property that allows a lens to work. Ordinary light beams, even if narrowly constricted or focused, become diffused and fuzzy over a distance because of interference among the various wavelengths of light contained in an ordinary light beam. Laser beams overcome this problem because they consist of only a single wavelength of light which can be narrowly focused into a thin beam that does not degenerate over distances.

Discussions of these and other examples may lead the group to a concept of "non-turning." Here are some typical ideas that students present:

- "It is obvious that: 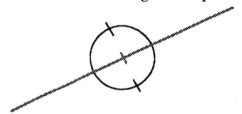 is not the shortest path between a and b."

- "If one turns one can form a triangle..."

- "There is a bend..."

- "A straight line is the most direct path..."

At this point, the class should try to arrive at a consensus about "not turning" as a good way to express going straight. There can be at least two expressions of "not turning": (1) going at a constant heading (the physical concept of sighting or light path); or, (2) having a 180° angle at any point on the path. The point in the discussion when someone introduces a 180° angle is an excellent time to get students to begin thinking about symmetry. The instructor should ask students: *"What is a 180° angle? What is it about a 180° angle that expresses "not turning"?"*

A 180° angle can be seen as half a circle. If two parts of a divided circle are congruent, then one has a 180° angle.

Now seems to be a good point in the discussion for the instructor to explicitly introduce the notion of symmetry. What symmetries does a straight line have? How do they fit with the examples that have come up in class and those mentioned above? Can we use any of the symmetries of a line to define straightness?

Give the students some time outside of class to think about and formulate some answers for these questions. It is a good idea to wait until after they have worked on Problem 1 to introduce the different symmetries. It is likely that the group will come up with all the symmetries that you would like to introduce. Even without symmetry, the students have more than enough to think about in order to solve the first homework problem.

Remind students that, in this course, they are the ones laying down the definitions, that nothing will be taken for granted, and that no answers are predetermined. The instructor is expecting to learn something **new** about straightness that will come from the students. Consequently, it is important that they persist in following their own ideas so that learning can happen for both the instructor and themselves.

When the time comes to introduce the symmetries of a line in class, it is preferable to ask students to introduce these by sharing their homework ideas with the rest of the group. In this way the instructor can convey to the students that homework is important for individual learning as well as for the development of ideas in the group. It stresses that the spirit of the course is about sharing ideas. Mathematicians and other practitioners share and compare ideas all of the time. Sharing ideas also helps to improve the quality of the homework, since students will know that they might have to explain their ideas to others. Most importantly, the act of explaining is a learning experience.

Example of Students' Work on Problem 1

Professor and T.A. comments:

Student answer:

Problem 1 - Karen D. Alfrey
1. The five year old sister of a friend of mine described a straight line as one that is not "messy."

A straight line always heads in the same direction. It never wiggles or turns.

If I am looking at a line "head-on" (as if it is coming out of my eyeball and going into the page) then if the line is straight I will only be able to see the point right in front of me. It will block the view of all other parts of the line:

my eye

straight line   not straight line

a) I check the straightness of my violin bridge by sighting along it. I also make sure the strings stretched over it are symmetrical by reflection through the bridge (i.e. that the bridge is perpendicular to the strings).

b) When building houses, we draw straight lines using a stretched string covered with chalk (a great way to draw long straight lines!). In drawing shorter lines (on a page, e.g.) I just try to move my pencil from start to finish without changing direction.

c) A straight line is symmetrical:
By reflection through a line perpendicular to it:

_____

By reflection through the line itself:

By half a rotation around its center:

d) A "straight line" is the one which traces a path between two points without swerving and covering the shortest possible distance.

Your definition of straight should be an expression of the path you made from a) to c).

Response to comments:

1.d) A straight path is one which does not turn and which possesses the following kinds of symmetry:

-reflection in the line:

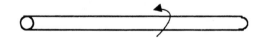

-reflection perpendicular to the line:

-half-turn symmetry:

-translational:

-rotation in 3-space:

-point reflection:

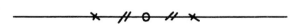

Chapter 2

Straightness on a Sphere

Students have begun to define straightness for lines in the plane. They used symmetry arguments to enforce intuitive ideas about what straightness really means. Now they will attempt to transfer those ideas to spherical surfaces. At first, it may be difficult for students to imagine the sphere in a way that considers only the surface. Intuition about properties of the plane is natural for people, in some sense, because it is so often the subject of mathematics lessons in formal education settings. An exploration of straightness on a sphere is a good way to start to think about some of the fundamental properties of a sphere, a region about which students may not have an immediate intuitive sense.

PROBLEM 2. *What is Straight on a Sphere?*

As they think about Problem 2, students should concentrate on the following ideas:

- Imagine being a bug crawling around on a sphere. The bug's universe is just the surface; it can neither fly off the sphere nor burrow into the sphere.

- What is "straight" for this bug? What will the bug see or experience as straight?

- How can you convince yourself of this?

Using the symmetry properties of straightness on the plane, students can convince themselves and construct arguments to convince others that the great circles on a sphere are the only curves that are straight with respect to the sphere. The instructor may want to introduce the terms *great circle* and *pole* in class.

At this point, students have the tools, through symmetry, to extend their notions of straightness to a sphere. The instructor can now ask them to think about symmetries of great circles, and this will lead students to compare the concept of straightness on a sphere to straightness on the plane. At this point, the following questions may be raised in class:

- Is there any other straight path on a sphere besides a great circle?

- Is a straight path always the shortest path on a sphere?

It is natural for students to have difficulties experiencing straightness on surfaces other than the plane. Consequently they will start to look at the curves on spheres as 3-D objects, and attempt to construct definitions and proofs with a 3-D notion in mind. Asking students to imagine being a 2-dimensional bug walking on a sphere emphasizes the importance of **experiencing** straightness and will help students to leave their limiting 3-D vision of the curves on a sphere. The instructor may want to ask students: 1) What is it that the bug has to do when walking on a sphere to be able to walk in a straight line? 2) How can the bug check to see if it is going straight?

Experimentation with models plays an important role here. Building and playing with models helps students to convince themselves that great circles are in fact the only straight lines on a sphere. In the process of becoming convinced of "great-circle-straightness," students will realize that there are commonalities between straight on the plane and straight on a sphere. They will then be ready to transfer, naturally, the symmetries of the line in the plane to the great circles on a sphere and, later, to geodesics on other surfaces. Students should have as many or all of the activities listed in their books. The instructor may want to organize some class time in which several of these investigations could occur, providing students with the appropriate materials when necessary. These explorations are crucial to a deep and meaningful understanding of straightness on the sphere. The instructor should not hesitate in stressing the importance of modeling.

Example of Student's Work on Problem 2

Professor and T.A. comments:

Student answer:

Problem 2 - Emma Lister

If the bug is walking a straight line on the sphere and the bug has walked completely around the sphere once, it will be back in the exact spot from which it started. If it continues to walk, it will retrace its exact steps.

Why?
What about latitudes?

If the bug is walking in a line that is not straight, once it has walked completely around the sphere, it will not be in the same place it started. Furthermore, if it continues to walk, eventually the path it is following will either cross itself or end up circling to a point at either pole.

Consequently, if the bug is walking in a straight path, it will see the exact same things each time around. To prove this, one can test a section of the line by cutting out a section of the sphere and flattening it out.

Is this possible?

On the flattened sphere section, if the line is straight by the properties talked about on question 1, then it is straight on the sphere. Therefore, this also excludes latitude lines from being straight on the sphere since they do not have a symmetric reflection about the line chosen perpendicular to a point on the latitude line.

Good!

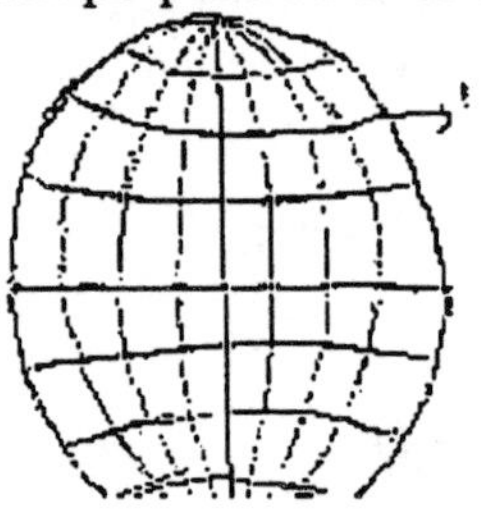

Actually, by drawing this picture, I see that this is not the reason.

The latitude line could be thought of as a straight line because, if a bug was to walk that line, it would end up in the exact same place after going around once.

Explain perpendicular

But, it seems that the latitude lines are not straight on the sphere because the same line is not perpendicular to every point on the line as it is with the equator. A longitude line is straight because the same line is perpendicular to its every point.

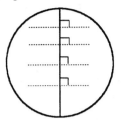

This shows that latitude lines and the equator are the only straight lines on the sphere for any pole you choose to pick. Hence, since these are the great circles, great circles are the only lines that are straight on the sphere.

Response to comments:

For my first response, I did not use a sphere as a model. I was relying on my two-dimensional drawings as my only visual aid. However, now that I have been using a soccer ball, I see things much more clearly.

While trying to explain the intrinsic curve of latitude lines, I stated that, if a line placed perpendicularly to a latitude line was moved along the latitude line across the sphere, the angle of the intersection would change due to the curve of the latitude line. However, using a soccer ball as a model, I found that idea to be completely incorrect. The perpendicular line moved across the sphere, does stay perpendicular to the latitude line as well as great circles.

So what makes a latitude line not straight with respect to the sphere? Well, I understand and accept all the reasons presented in class on Thursday (9/3) except one. One person suggested that, if a sphere was covered in paint and rolled on a flat surface, the line that would be made would be a great circle.

If you paint the soccer ball along a latitude line and roll it on the floor, you will not get a straight line. Try it!

latitude line   **great circle**

Painted Area

Well, it is correct that the center of this line would be a great circle. But, depending on the width of the painted line, and depending on the amount of pressure placed on the ball as it is being rolled, there is a varying amount of painted area surrounding the center of this line which could be regarded as latitude lines. Therefore, what is defined as a great circle must be defined by width in order for one to follow this logic.

This also relates to the problem regarding width as discussed in another model. [*She is referring to the ribbon test.*]

It was mentioned that a bug walking along a latitude line would take shorter steps with one foot than with the other. If he started at any point, in any direction, and walked with exactly equal steps with both feet, he would be walking in a straight line with respect to the sphere and, therefore, a great circle. Therefore, the bug should be able to tell the difference when he is walking on a curved latitude line. This is dependent on the size of the bug relative to the size of the sphere.

What about the legs of the bug? Wouldn't the inner legs travel less than the outer legs?

If the bug is big enough, and the sphere is small enough, the bug will be aware of his curved path. But if the bug on the sphere is comparable to the size of humans on earth, then, even if he is walking along a latitude line he will think he is walking in a straight line.

<u>Symmetries</u>:

Because the latitude lines are parallel to the equator, which is a great circle the symmetries of latitude lines and great circles are, in fact, similar.

1. Both great circles and latitude lines have reflection symmetry perpendicular to the line.

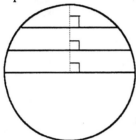 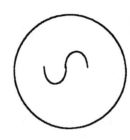

On the sphere, this has central (1/2 turn) symmetry, but doesn't have it in 3-space in the sphere because of the curve of the sphere.

2. Consequently, when the lines are viewed as such (see figure above) they all appear to have half-turn symmetry also.

3. Latitude lines and great circles have translation symmetry.

4. Great circles have reflection symmetry in the line.

You are again flattening your sphere and that does not represent it accurately.

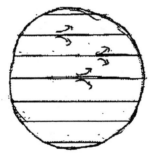

Viewing the sphere this way, it appears that latitude lines also have reflection symmetry in the line. Here the latitude lines appear straight:

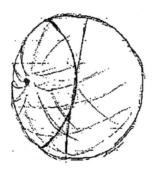

However, viewing the sphere from this way, it appears that great circles have reflection symmetry in the line and latitude circles <u>do not</u>.

Here the latitude line appears curved.

Neither (or both), use a ball. Note <u>any</u> circle looks straight edge-on.

I am a little confused as to which is the correct way of seeing this with respect to the sphere.

Chapter 3

What is an Angle?

Students have spent some time looking at lines without necessarily thinking about the shapes they might form or the angles in which they meet. Again, students are being asked to question some fundamental assumptions about a group of objects on the plane — angles. Students may want to think about Problem 4 before Problem 3, and instructors should encourage multiple approaches, including looking at the problems out of order. Students will inevitably revisit problems they think they have "solved" after a new idea comes up in class, or in the course of investigating subsequent problems. So, doing problems out of order may give students a chance to anticipate the direction in which they would like to build up their definitions and proofs.

PROBLEM 3. *Vertical Angle Theorem (VAT)*

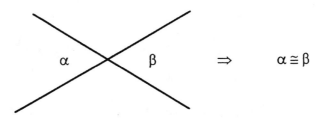

Prove: A pair of opposite angles formed by two intersecting straight lines are congruent.

We do not have in mind a formal two-column proof as used in high school geometry. Mathematicians in actual practice usually use a proof to mean a discussion sufficient to convince any reasonable skeptic. [Hint: Show how you would "move" α to make it coincide with β. What symmetry properties of straight lines are you using?]

Does your proof also work on a sphere?

PROBLEM 4. *What is an Angle?*

Give some possible definitions of the term "Angle." Do all of these definitions apply to both the plane and the sphere? What are the advantages and disadvantages of each?

What does it mean for two angles to be congruent? How can we check?

Whatever proof students present for the Vertical Angle Theorem, that proof has a "built in" notion of angle. In general, the purpose of introducing Problem 3 before Problem 4 is to have students prove the Vertical Angle Theorem and decide what unstated, but assumed, definition of an angle they were using in their proof. Then, with this knowledge, students can launch into a discussion of the many different angle definitions that are possible. Some students may give answers to the two problems that are independent of one another, and so the instructor should emphasize the importance of doing both problems together.

The instructor should also emphasize, once more, the importance of thinking locally. The notion of angle depends only on what happens at the turning point. If one draws an angle, and starts to erase both legs simultaneously without erasing the corner, the angle is still there.

Students will come up with many different proofs, and it is important that they share those proofs with the group. Questions raised by peers and the instructor give students a chance to strengthen their arguments. Also, having students informally present their proofs to the class is a good opportunity for the instructor to call attention to the different notions of angle that are implied by the different proofs.

The three proofs given in the student manual are based on different perspectives of angle. The first assumes a measure-based definition of angle; the second demonstrates angle as a rotation; and the third proof uses a transformation to show that an angle is a geometric figure. We present them here to illustrate some ways that the instructor can look at the proofs students will present in the homework.

Example of Students' Work on Problems 3 and 4

Professor   and   T.A.   Student answer:
comments:

Problems 3 and 4 - Joanne Galinsky

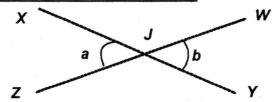

The two ideas that immediately spring to mind to show that *a* is congruent to *b* are: (1) fold *a* onto *b* and (2) rotate *a* to coincide with *b*.

By doing the re-flections one at a time, haven't you destroyed the angles?

(1) Using the reflection symmetry of a line through a perpendicular line, we can show that *a* is congruent to *b*. Let's assume that the point *J* is the midpoint of both line *XY* and *ZW*. If we reflect line *XY* through a line perpendicular to it at *J*, point *X* will lie where *Y* is and vice-versa. The same holds for line *ZW*. So the two angles must be congruent; we now have angle *XJW* where *YJW* used to be and vice-versa.

Actually, I just realized that this is not the same as folding the paper at all. To fold *a* onto *b*, we would have to make the fold along the line which bisects angles *XJW* and *ZJY*. *XJ* would lie on top of *WJ*, and *ZJ* would lie on top of *YJ*, which is not what I described above. At the moment I'm not sure what property of straight lines I'm using to prove this; I just know it works because I tried it!

Why? —————→

!

Response:

    Fold so $x \rightarrow w$

If the top matches, the bottom must match because the line stays straight.

(2) Rotating *a* onto *b* utilizes the half-rotation symmetry of straight lines. A half-rotation of *XJ* would make it lie on top of *YJ*; the angle between them, *a*, would be preserved, since we rotated both segments by the same amount and must be congruent to *b* since it now lies on top of it.

good

Do these methods work on the sphere? Yes, with a few assumptions and minor changes.

(1) We can't "fold" anything on the sphere, but using reflection symmetry through a perpendicular line works as described above as long as we assume that our "lines" are <u>constructed of something pliable, like string</u>. Otherwise the reflection would curve up away from the sphere instead of onto the sphere.

Use mirror

During the line actions of symmetry, the properties of the geodesics have to be maintained.

(2) As far as I can tell, rotating works just fine on the sphere. We do not have to deal with any curvature problems because rotating never takes the lines off the surface of the sphere.

Response:

    Rotate around the bisector — same as mirror which works on sphere.

Good

Problem 4. What is an angle? When are two angles congruent?

My first thought was that an "angle" is the antonym to "straight line", but this is not quite right. Lines that are not straight do not necessarily have to be angles; they can be smooth curves or irregular curves. Then I thought an angle could be considered more accurately as two intersecting straight lines. Then, on a sphere the only angles are the intersections of great circles; for example, the intersection of a longitude line and latitude line would not be considered an angle.

Whereas, for a straight line the concept of "not turning" was important, for the angles there is a concept of turning that must be considered. This turning cannot be a gradual curve; it must be a sharp turn on a single point in space. To walk the path of an angle, I would have to first walk straight in one direction, then STOP; turn a certain amount; and follow a new straight path. As a clarification, I must add that, since a straight line is an angle, it is not necessary to TURN when you stop, i.e., you can turn 0 degrees if you wish. This works in Euclidean space and on the sphere. On the sphere I am, in essence, following the path of one great circle; stopping and turning at an arbitrary "pole"; and then following the path of another great circle.

One way to check if two angles are congruent is to "pick up" the first angle and see if it fits perfectly over the second angle. We can use this all the time in Euclidean space as long as we allow ourselves to flip the

Not if you use a mirror instead of flipping.

angle over if necessary. On a sphere this method works fine as long as the two angles are identical rather than mirror images of each other; with mirror images, we'd have the problem of curvature — if we tried to place one on top of the other, they'd curve away from each other. I suppose if we were allowed to flatten them, we then could see that they were congruent. If we could place a mirror between the two angles and somehow project the image from the mirror onto the other side of it, the image of the angle would coincide with the angle on the sphere, thereby showing that they were congruent.

Can you always slide one angle into the other without leaving the sphere?

Response to comments:

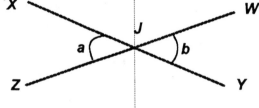

Here are a few ways to think about this:

(1) (I know you don't like this one, but I will try to explain as convincingly as possible.) Using the reflection symmetry of the line through a perpendicular line, we can show that a is congruent to b. If we reflect line XY through a line perpendicular to it at J, point X will lie where Y is and vice-versa. The same holds for line ZW. So the two angles must be congruent; we now have an angle XJZ where YJW used to be, and vice-versa.

The way I view an angle is that it doesn't have to always "be there" or remain intact while you are playing with the lines to show one angle is congruent to the next. If you were to rotate the angle to prove congruence, it wouldn't matter if you rotated both lines of the angle at once. or if you rotated them at different times. If both lines are rotated the same amount, regardless of when, the end result will be that the original angle will appear at a new orientation dictated by how far you rotated the lines. The key is not to do something to the lines at once;

So are you attrib-uting to reflection action a similar property? Explain it !

rather, it is to do the SAME THING to both lines. If we perpendicularly reflect each line, we have done the same thing to each line, thus preserving the angle between the lines. So when we are done with this process, the original angle will reappear in the new location.

(2) To fold a onto b, we would have to make the fold along the line which bisects angle XJW. XJ would lie on top of WJ because of the definition of an angle bisector—the lines forming the bisected angle are bilaterally symmetric. Claiming that ZJ would lie on top of YJ is a bit trickier, because you cannot assume that we have bisected ZJW (since the vertical angle theorem is what we are trying to prove). The way I see it, the bottom segments must match up because the lines XY and ZW stay straight, so what happens to one side happens to the other. To illustrate, let us assume we are rotating the lines around the bisector instead of folding it, or we are placing a mirror along the bisector (since the same result occurs). Then we can see more clearly that the lines stay straight during the process, and what happens to one side of the line, is also happening to the other side, so XJ must lie on top of WJ and ZJ on top of YJ. The mirror method also works on the sphere.

3) Rotating a onto b utilizes the half-rotation symmetry of straight lines. A half-rotation of the line segment ZJ would make it lie on top of WJ; a half-rotation of XJ would make it lie on top of YJ; and the angle between them, a, would be preserved since we rotated both segments by the same amount, and a must be congruent to b since it now lies on top of it.

Do these methods work on the sphere? Yes.

(1) Consider two great circles intersecting at an arbitrary pole. By using a mirror, we can reflect each line onto itself, and the lines will lie on top of the lines on the other side of the mirror. (This works fine, assuming you believe my argument in 2-space.)

(2) We cannot "fold" anything on a sphere, but using reflection symmetry through a perpendicular line works if we use a mirror to illustrate it. The reflection of an angle in the mirror produces a vertical angle if we place the mirror along the bisector of one of the supplementary angles to a or b.

(3) As far as I can tell, rotating works just fine on the sphere. We don't have to deal with any curvature problems because rotating never takes the lines off the surface of the sphere.

Problem 4. What is an angle? When are two angles congruent?
One way to check if two angles are congruent is to "pick up" the first angle and see if it fits perfectly over the second angle. We can use this all the time in Euclidean space as long as we allow ourselves to flip the angle over if necessary. On a sphere, it seems really easy. If you

allow yourself to slide an angle over the sphere's surface, you can orient it in some way so it coincides with a congruent angle. This took me awhile to see. At first I was considering triangles rather than angles. For example, you can't slide a right triangle around a sphere to make it coincide with a mirror image of itself. But that has to do with LENGTHS of the sides of the triangles rather than the ANGLES, which are of differential size, or local. We can isolate the right angle of a triangle, and slide that around to coincide with the right angle of the other triangle, showing that the two angles are congruent, just as we can with any other angle.

Chapter 4

Straightness on Cylinder and Cone

In Problem 2, students are asked to transport the symmetries of a line in the plane to a sphere. This exercise encourages them to consider straightness from the bug's point of view. In particular, students should see that it is impossible to transport the central symmetry property to the sphere. The questions in Problem 5 also encourage students to consider straightness from the bug's point of view. They are similar to the questions in Problem 2, but this time the surfaces are a cylinder and a cone.

PROBLEM 5. *Intrinsic Straight Lines on Cones and Cylinders*

What lines are straight with respect to the surface of a cone or a cylinder? Why? Why not?

Geodesics on Cylinders

Initially, students will make empirical observations about the ways in which the bug might travel in a straight path on a cone and a cylinder. It is important to convey to students that, even though those observations may not appear to be mathematically justified, they are important and should be shared with the group. Some of these preliminary findings might include the classes of straight lines on a cylinder and a cone. First, let's look at the three classes of straight lines on a cylinder.

Straight line path made by a vertical generator:

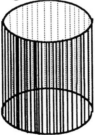

Path made by the intersection of a horizontal plane with the cylinder, a *great circle* or a *generator circle*:

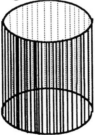

A path which makes a spiral or helix of constant slope around the cylinder:

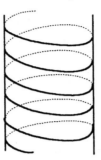

Be sure to ask students to explain how they decided these lines were straight: What methods and what experiments did they use? Did they predict that these lines would be straight or were they surprised when they found them? It is important that students explain how they constructed these straight lines.

Geodesics on Cones

Now, let's look at the classes of straight lines on a cone.

Path made by a generator: As students consider the straight path made by a generator on a cone, they will have to make a decision about straightness at the cone point. Students might decide that there is no way for the bug to find out how to go straight from the cone point, and thus the straight path ends there. Or, they might decide that the bug has some knowledge of its world and knows the cone angle. In this second case, when the bug is at the top, it can go straight down by bisecting the cone angle (see next figure). It is a good idea to remind students that symmetries may be very useful in deciding what it means to walk straight from the cone point. Some students will say that the line ends at the cone point because some of the straight line symmetries do not hold there; e.g. translation symmetry. Other students will choose to continue the line on the other side of the cone, dividing the cone into two symmetric parts. This path has most of the symmetries of the straight line at the cone point. (Which ones?)

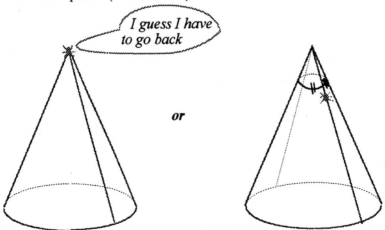

Straightness at the Cone Point

Path straight and around: Students will have different perspectives on the kind of geodesic on a cone that goes straight around the cone. Some students will say that this kind of geodesic wraps around an infinite number of times. Others will say that a straight-around geodesic seems to wrap around once, intersect once, and then go straight down on both ends without intersecting again. Other students will say that these geodesics never intersect even though they are not generators. Obviously, students will need to work with cones of varying angle sizes. The instructor should encourage students to bring their models to class and to explain their conjectures to the rest of the group. Soon enough, students will realize that their conjectures depend upon the size of the cone angle, and the group can state their findings. The maximum number of times that a geodesic on a cone can intersect itself can be summarized in the table below. This data is preliminary in nature. Later in this chapter we will refine and justify the numbers in this table by using covering spaces.

Cone Angle in Degrees	# of Self-Intersections
0-59°	at least three times
60-89°	twice
90-179°	once
180-360° (360° = plane)	none
More than 360°	none

Local Intrinsic Symmetries

By the end of the first set of suggestions, students should realize that when a piece of paper is rolled or bent into a cylinder or cone, the bug's local and intrinsic experience of the surface does not change except at the cone point. Extrinsically, the piece of paper and the cone are different, but in terms of the local geometry intrinsic to the surface they only differ at the cone point. Introduce the idea of isometry in this context and be sure students feel comfortable talking about the cone, cylinder, and plane as *locally isometric* surfaces.

PROBLEM 5a. *Covering Spaces and Global Properties of Geodesics*

How many times can a geodesic intersect itself? How are the self-intersections related to the cone angle? How can we justify this relationship? How do we determine the different geodesics connecting two points? How many are there? How does it depend on the cone angle? Is there always at least one geodesic joining each pair of points? How can we justify our conjectures? With what kind of tool could we equip the bug to help with this investigation?

In Problem 5a, students are provided with a new tool for exploring geodesics on a cone and on a cylinder, *(n-Sheeted Covering Spaces)*. These allow further exploration of properties of geodesics on cones and cylinders, as well as allow students to justify the empirical observations they made solving the first part of Problem 5.

Different Geodesics Joining Two Points

There are several different kinds of geodesics joining two points. Here are a few.

On a Cylinder: We can see that there are an indefinite number of lines joining two points on a cylinder if the two points are not on the same great circle. Suppose we have two points *a* and *b*, and we unroll the cylinder into an *n-sheeted* cover. There will be a lift (copy) of *a* and a lift of *b* on each sheet. Pick one of the lifts of *a* and join it to each of the lifts of *b* by lines which are straight on the covering. The images of these *n* lines will be *n* different geodesics on the cylinder.

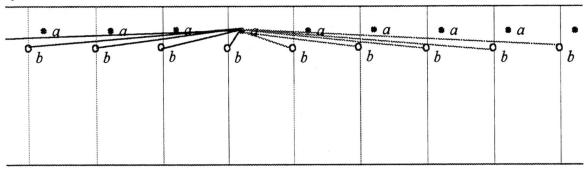

Thus we can construct as many geodesics as we wish joining two points on the cylinder.

On a Cone: It is more complicated to compute the number of lines that connect two points on a cone. The number will depend upon the cone angle and the location of the points that one wants to connect. Given a ϕ angle cone, let us construct covering sheets for that cone until the sheets cover the plane (this is once around). Given two points on the cone, mark the lifts of those points on the sheets of the covering space considered. Now join one of the lifts of *A* (*A'* in the figure below) to all the lifts of *B*. These lines are all different ways of connecting these two points on the cone.

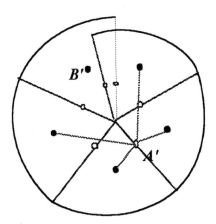

Contrary to what some students may predict, the number of lines connecting two points on a cone with cone angle ϕ is finite; there are approximately 360°/ϕ lines, which is the number of sheets of the covering that produce 360° at the lift of the cone point. Note that in the figure, the lift of *B*, represented by *B'*, cannot be connected to *A'* by a straight line, thus the relative positionings of *A* and *B*, together with the cone angle, will influence how many geodesics join the two points. Adding more sheets to the covering will not result in more geodesics joining *A* to *B*.

Initially, with the addition of coverings, students will feel sure that more geodesics are being added. Continually encourage students to make models as they make and confirm conjectures.

Let us see an example where there are no straight lines connecting two particular points. Consider a cone with angle 450°, and two points *a* and *b* such as those in the figure below. In this case there is no geodesic that connects the two points.

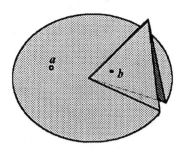

No geodesic can connect *a* and *b*. Try it!

Number of Times a Geodesic Intersects Itself

The number of times that a geodesic can intersect itself depends on the cone angle. Students should become comfortable using covering spaces, specifically here to estimate the maximum number of self-intersections on a cone with cone angle ϕ.

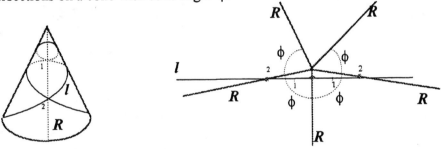

Draw a straight line *l* on the cone. Consider the ray *R* such that the line *l* is perpendicular to it. The lift of the geodesic *l* (this is the corresponding geodesic on the cone) will cross the ray *R* exactly once for every time its lift crosses the seam of the covering. Note that the seams between individual wedges are lifts of the ray *R*. The number of times that the geodesic can cross the ray depends upon the number of times the cone angle ϕ fits into 180°. The geodesic goes around the cone once for every time it intersects the seam. And every time it goes around the cone, it intersects itself. A geodesic will intersect itself every time it crosses the ray *R*, and will also intersect again in the back of the cone opposite to *R*. Thus, the number of intersections is roughly the number of times that the cone angle fits into 180°. Consequently, on a cone with an angle greater than 180°, geodesics never intersect themselves.

On a cylinder, with the exception of great circles, geodesics do not intersect themselves.

At this point, the students should be thinking intrinsically about the sphere, cylinder, and cone surfaces. In the problems to come, they will have opportunities to apply their intrinsic thinking when they make their own definitions for triangle on the different surfaces and investigate congruence properties of triangles.

Connections with Differential Geometry

Covering spaces bring out the differences between the three surfaces in a very natural way. Students should realize several different features of the surfaces. For example, locally and intrinsically, a cylinder is the same as a plane. The cone is locally indistinguishable from a plane if one is excluding neighborhoods that include the cone point. However, the cone point sets the cone apart from the plane and cylinder, and aligns it closer to the sphere.

Depending on the background and interests of your students, you may wish to include here more connections with differential geometry and topology in this discussion. It would be particularly appropriate to introduce the hyperbolic plane in a form that is accessible to your students.

Example of Students' Work on Problem 5

Professor and T.A.   Student answer:
Comments:

Problem 5 - Suzy MacDaniel

A straight line for a cylinder: (these are straight according to the bug)
    Several types:
What it looks like on a plane:

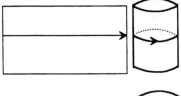

Type A: (Made by cutting a plane perpendicular to the axis of the cylinder.) It is actually a circle.

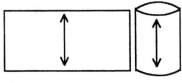

Type B: Parallel to the axis:

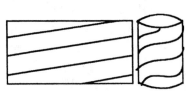

Type C: Barber shop lines. These lines twirl around. If a bug were to walk in a direction that was not perpendicular to the axis, or parallel to the axis he/she would walk in one of these barber shop lines and never cross his/her path.

This direction can vary relative to the perpendicular and parallel type of lines. Ex:

Do they look like that if you draw them on a real cylinder?

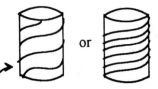

or

In this case, the bug would be crawling in a direction close to Type A direction.

The bug crawls in a direction close to the Type B line.

A straight line for a cone:

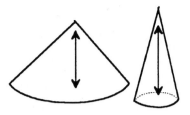

Type A:
An extrinsically straight line which can be in the same plane as the axis.

Good!!

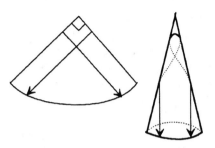

Type B:
There are some paths on which a bug can crawl where it will cross its path once. I found these lines to be two intersecting lines with 90° between them (on a flat piece of paper).

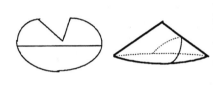

Type C:
I found straight paths on the cone where they never intersect on a flat piece of paper. They were perpendicular to each other. I usually found these on wider angle cones, i.e. >180°.

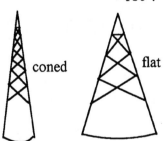

coned   flat

Nice!

I noticed that, as I decreased the cone angle, the number of intersections occurred among the line:

Does this mean that, as the cone gets infinitely smaller, the number of intersections gets infinitely larger?

According to a bug, there is no change in straightness, even when a flat surface is rolled into a cylinder, or when a flat surface is rolled into a cone.

Or, imagine taking a flat piece of paper with a slit in it and gradually manipulate it to turn into a cone:

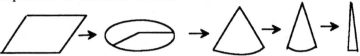

Why?

I believe that bug would not see a difference in straightness.

Response:
Intrinsically, there is no change in the surface to the bug. But, if there were a change that the bug could notice, I could notice that change, too. Or rather, an intrinsic change is also an extrinsic change. But an extrinsic change is not always an intrinsic change.

If a cone is intersected by a plane perpendicular to its axis, a curve is formed:

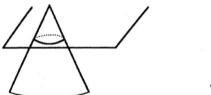

A geodesic will never intersect itself with respect to the cylinder. It will coincide with itself if it is perpendicular to the axis.

Can it be more than one geodesic joining two points?

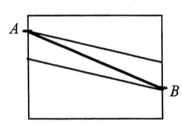

On a cylinder yes. If you rolled this up, you would see that *A* and *B* are joined by two geodesics. In fact there are an infinite number of geodesics that join *A* and *B*. Any slope [-1/2, +1/2] would work.

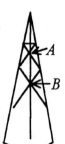

For a cone: There is always one geodesic that will join two points.

I just changed my mind. There are usually two geodesics for each two points. One goes in front and the other goes in back.

Nice drawings and models!!

Response to comments:

From several scratch drawings I came up with some formulas. (Please see the attached notebook paper for my pictures [*not included here*] which stand behind these algebraic formulas.)

Given the cone and its angle ($x°$) you can find the number of geodesics that connect two given points a and b which are on the cone.

What if $x > 360°$?

Case 1: n = integer; $n = 180 / x$; g = number of geodesics

$2(180 / x) - 1 + 1$ (for the geodesic that touches the vertex) $= g$
therefore, $2(180 / x) = g$

a and b and the cone point might not be collinear

Case 2: n is not an integer and [*Here* $[n]$ = integer part of n.]
$(n - [n]) > ($ radial distance between a and $b) /$ cone angle

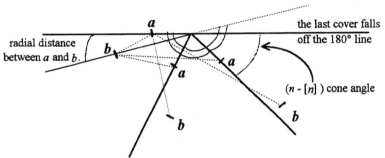

In this particular case, the cone angle is about $75°$ thus $180 / 75 = 2.4$, but $.4 > (15 / 75) = .2$. So the following formula applies in this case: $2 ([n] + 1) = g$. In this particular case, $2(3) = 6 = g$ (one geodesic is for the one that touches the vertex, 5 others are dashed)

Case 3: n is not an integer and
$(n - [n]) \leq ($ radial distance between a and $b) /$ cone angle

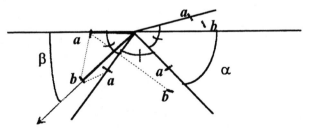

In this case the angle α is less than the angle β (radial distance) and so the following formula applies:

$$g = 2[n], \text{ where } n = 180 / \text{(cone angle)}.$$

In this particular case, $g = 2(2) = 4$, that is, 3 normal geodesics and 1 for the vertex (dark in figure above)

Nice Discussion!

Say the cone angle is 270°. I will test my formula:
$$n = 180 / 270 = .66...\ .$$

Then n is not an integer; it is case 2, therefore $g = 2([n] + 1)$. We have 2 geodesics, in figure: one light and one dark.

If radial distance between a and b is less than the left-over angle from 360 / cone angle, then the formula is $[(720 / \text{cone angle})] - 1 = n$. But if the radial distance between a and b is greater or equal to the left-over angle, then the equation is: $[(720 / \text{cone angle})] - 2 = n$.

What if the cone angle is bigger than 360? Say 450°, but for 450° there are two cases depending on where A and B are located on the cone:

(1) there is only one
(2) there are none.

Please refer to the diagram on page 45 [*Figure 4.20 in main text*]. There are no geodesics between A and B but there is one between nearby points.

So, if the shortest of the two radial distances between A and B is less than 180° for a cone with cone angle greater than 360°, there is one and only one geodesic connecting A and B. However, if the shortest radial distance is greater than 180°, then there are zero geodesics joining two points.

Aren't these already counted?

So, the number of geodesics that can join two points on a cone depends on the cone angle. For a 45° cone angle the number of geodesics that can be drawn between two points on the cone is:
$$8(A \text{ to } B) + 7 = 15.$$

This is a geodesic from B to B

For angle $\alpha = 120°$, $n = 5$:

Model a 60° cone and draw all possible

So: $(360 / \alpha) + (360 / \alpha) - 1 = (720 / \alpha) - 1 = n$.
For $\alpha = 60$ we have: $(720 / 60) - 1 = n$; $11 = n$.

But, what if A and B are located on the same geodesic that goes

through the cone point? Like this:

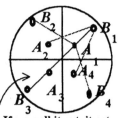

Normally there would be
$(720 / 90) - 1 = n;\ 7 = n.$

If you roll it out, it gets weird because of this geodesic

I think you are working on the formulas away from your models

If you count the geodesics you will count only 5. This is because the geodesic that joins A_1 to B_1 joins also B_1 to A_3 and A_1 to B_3. The formula for a cone when the points are on the same geodesic is:
$$(720 / \alpha) - 3 = n.$$
If α does not go into 360° evenly, then there is a change in the formula.

Test out my formula:
$\alpha = 20°;\ n\alpha = 180;\ 20n = 180;\ n = 9$ or $\alpha(n + 1) = 180;$
$20(n + 1) = 180;\ n = 8.$

$\alpha = 70°;\ (n + 1)70 = 180;\ n + 1 = 2 + (4 / 7);\ n = 1 + (4 / 7) = 2$

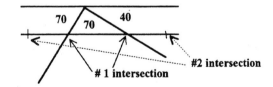

I never get it straight...

$\alpha = 450°;\ n + 1 = 180 / 450 = 2 / 5;\ n = -3 / 5.$ The formula still works since you are supposed to go to the next highest integer. What kind of function is that? Med? trunc?

Response to comments:
To calculate the number of intersections that a geodesic has given the cone angle you could use the following formula: $n\alpha < 180°$ or $(n + 1)\alpha = 180°,$ where α is the cone angle and n is the # of times a geodesic intersects itself.

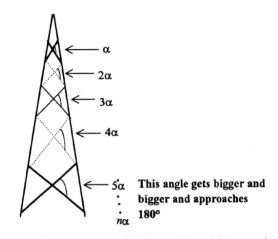

This angle gets bigger and
bigger and approaches
180°

For a 180° cone: $n(180) + 180 = 180$; $n = 0$.

For a 179° cone: $179n + 179 = 180$; $179n = 1$; $n = 1$. Always round n up to the next highest integer if it is not an integer.

I think only the case of the cones with angles >360°, unless I missed them...

I have looked over #5 and I'm not sure what I have left out. Can you point out to me what is left?

Response to comments:

When a cone angle is greater or equal than 360° the number of intersections of a geodesic is 0.

How many geodesics joining two points? Why? How do you see it?

Response to comments:

For cone angles greater than 360 you have 3 cases. You must first take the angle measurements of the two points by drawing a geodesic from the vertex to point A, then another to point B (from the vertex!). Then you have formed two angles: call them α and β; $\alpha + \beta$ = cone angle:

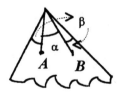

If $\alpha > 180°$ and $\beta > 180°$, the only geodesic joining A and B is through the vertex.

If $\alpha < 180°$ and $\beta > 180°$ (note that β must be greater than $360° - \alpha$), then there is one geodesic that joins A and B that does not go through the vertex. This geodesic passes through the angle α.

If $\alpha > 180°$ and $\beta < 180°$, a similar case holds.

Chapter 5

SAS and ASA

Problems 6 and 7 ask students to prove theorems that can be seen as dual to one another in the sense that their validity depends on two familiar properties of the plane:

· *Two points determine a unique line segment joining them.*

· *Two intersecting lines intersect in a unique point.*

PROBLEM 6. *Side-Angle-Side (SAS)*

Are two triangles congruent if two sides and the included angle of one are congruent to two sides and the included angle of the other?

In some textbooks, SAS is listed as an axiom; in others, it is listed as the definition of congruency of triangles, and in still others as a theorem to be proved. Regardless of how one considers SAS, it still makes sense and is important to ask: Why is SAS true on the plane? One can also ask: Is SAS true on spheres, cylinders, and cones?

When students look at why SAS is true on the plane; they will use the property of straight lines on the plane to determine that there is a unique straight line segment joining two distinct points. The figure below illustrates how this property can be used to prove SAS on the plane. Suppose that $\triangle ABC$ and $\triangle A'B'C'$ are two triangles such that $AB \cong A'B'$, $\angle BAC \cong \angle B'A'C'$ and $AC \cong A'C'$. Translate $\triangle A'B'C'$ along AA' so that A' coincides with A. Since the sides AC and $A'C'$ are congruent, we can now rotate $\triangle A'B'C'$ (about A) until C' coincides with C. Given that *on the plane there is only one straight line segment joining two points*, if B' is now coinciding with B, then the third sides (BC and $B'C'$) will also coincide. Thus, in this case, the two triangles are directly congruent. On the other hand, if B and B' are not coincident after rotating AC' to AC, then a reflection (about $AC = A'C'$) will complete the process and show that $\triangle ABC$ is congruent to $\triangle A'B'C'$. This proof is fairly standard, and many students will come up with similar constructions. As always, encourage students to try something new, to follow their intuition even if it seems to contradict what they learned in high school geometry, and above all, to draw pictures and make models.

Step 1 - Two Triangles with SAS

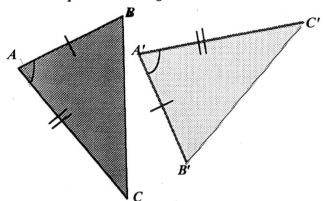

Step 2 - Translate A'B'C' along A'A

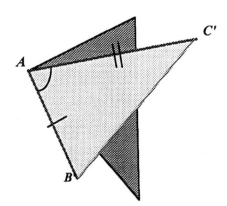

Step 3 - Rotate A'B'C' about A

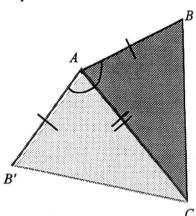

Step 4 - Reflect about AC

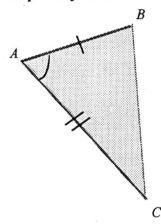

Encourage students to look at triangles for which SAS is not true. They will see some of the consequences of the properties of geodesics on spheres, cones, and cylinders. They will look closely at the features of triangles on those surfaces, and their findings should challenge their expectations about what a triangle is. These expectations go beyond what can be put into words as a definition of triangle. When students look for a definition of a triangle for which SAS will hold on these surfaces, they will try to stay close to their intuitive notion of a triangle. Students usually come up with examples of very strange triangles. For example, refer to the following diagram:

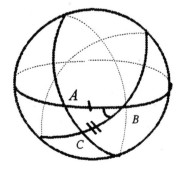

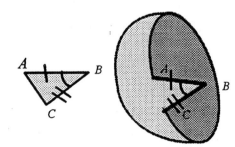

Two Choices for SAS on a Sphere

When students construct the above example, they may decide to accept the smaller triangle in their definition of triangle, but to exclude the larger triangle from their definition.

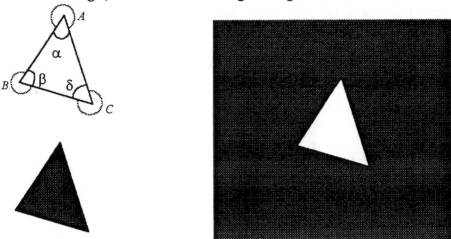

On the plane, what we want to call a triangle has all of its angles on the inside. Also, there is a clear choice for *inside* on the plane; it is the side that has finite area. The restriction that the area on the inside has to be finite doesn't work for spherical triangles (see figure above) because all areas on a sphere are finite. So what is it about the large triangle that challenges our view of triangle? Some students might try to resolve the problem by specifying that each side must be the shortest geodesic between the endpoints. However, antipodal points (that is, a pair of points that are at diametrically opposite poles) on a sphere do not have a unique, short geodesic joining them. On a cylinder we can have a triangle for which all the sides are the shortest possible, yet the triangles do not satisfy SAS. Let's look at this example:

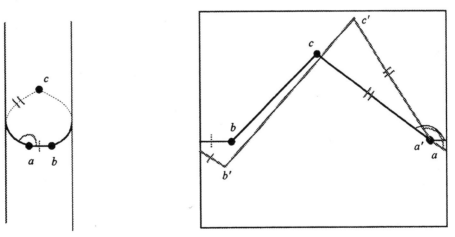

Similar examples can be constructed on cones by including the cone point inside of at least one of the triangles. From the top, a triangle that encircles the conepoint looks more like what we think of as a triangle than the triangles on the cylinder in the example above. In addition, when faced with the decision as to what the interior of a triangle on a cone is, one choice will always have a finite area. Suggest to students that covering spaces can help them in their investigations of such triangles. For example, what happens when we try to unwrap or lift one of these triangles onto a covering space?

Can one find a collection of "small" triangles for which SAS is true for spheres, cylinders and cones? Can one find a common argument for the small triangles on all the three surfaces?

These questions prod the students to make their own definitions. Students are not asked to make the strongest definitions possible (*strongest* in the sense of getting the strongest theorem possible), but to find a class of small triangles that makes sense to them on each of the surfaces **and to prove** (give a convincing argument) that for these small triangles SAS is true.

Let's look at some of the different definitions of a triangle that students might conceive in their attempts to make SAS hold on the three surfaces. The instructor should keep in mind that these examples should come from the students; the purpose of illustrating them here is to exemplify the diversity of ways in which the students might look at the counterexamples referred to above. Even though some of the definitions might be logically equivalent, they certainly are not equivalent in the images they point to, the meanings they emphasize, or the subjectivity involved in the way they were articulated. All of the definitions assume the following basic properties of triangles. It is likely that these properties will be naturally mentioned by the students in class discussion:

- *A triangle has three vertices that are pairwise joined by geodesic segments. These segments are called the sides of the "triangle."*

- *A triangle has three angles formed by the three pairs of sides.*

- *All of the angles are on the same side in the sense that if one walks along the sides of the triangle in a complete circuit, then each angle is either on the right-hand side or each is on the left-hand side.*

Now we turn to the definitions of small triangles that students have proposed, followed by proofs that those definitions suffice to prove that *SAS is true for small triangles*. Note: To conserve space we have edited the students' original proofs, but in doing so, we have not changed any of their original arguments.

Definition 1

A triangle on a sphere is a small triangle if it is contained in an open hemisphere. On a cone or cylinder a triangle is small if it is contained in an open one-sheeted covering space. (Here, *open* denotes that we are not including the great circle that is the boundary of the hemisphere.)

Proof:

Students will have various ways of seeing that in an open hemisphere or open one-sheeted cover every pair of points can be joined by only one geodesic segment, and thus the planar proof for SAS holds. Note that on the cone and cylinder the above restriction makes all the triangles intrinsically the same as plane triangles.

Definition 2

A small triangle on a sphere is a triangle for which all its angles are less than 180°.

Proof:

This definition makes SAS true on a sphere, but not on a cylinder or cone. Let us see how this definition works on a sphere. The two triangles use translation, rotation, and reflection to coincide with the two given sides and the included angle. Since the Angle in Side-Angle-Side is not 180° the remaining two vertices are not antipodal, and so they can be connected on a sphere in only two ways. Since at each vertex the angles formed by these two ways of connecting must differ by 180°, at most one of them produces only angles less than 180°. Thus again the planar proof of SAS applies.

The following figure depicts a triangle which shows that one must say more on the cylinder than just that the angles are less than 180°:

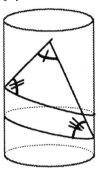

Definition 3

A small triangle on a sphere is a triangle for which each side is strictly less than half of a great circle.

Proof:

This definition resolves the problem of having more than one geodesic connecting two points in a different way. This is because two great circles intersect at antipodal points. If a pair of points has a geodesic segment of length less than half a great circle joining them, there is no other such segment. Thus again, the planar argument can be applied.

Definition 4

A triangle on a sphere is a small triangle if it is not intersected by every great circle. On a cylinder or cone, a triangle is small if it is not intersected by every vertical generator.

Proof:

If the triangle is small on a sphere then it is not intersected by one great circle and thus the triangle must lie in one of the two open hemispheres determined by that great circle. On a cylinder or cone, there is one vertical generator that does not touch the triangle. Cut the surface along that generator and flatten it. The triangle will be a plane triangle for which SAS holds.

Definition 5

A small triangle on a sphere is a triangle that cannot contain a pair of antipodal points. On a cylinder and cone, the triangles cannot contain the tube of the cylinder or the cone point.

Proof:

If the triangle does not contain any pair of opposite poles, it is not possible to connect the unconnected vertices by a path equal to or longer than half a great circle. This requirement makes the path between two points unique. On a cone, we also exclude from the class of triangles those which include the cone point in order to avoid a case similar to that shown above, in which the triangles include the tube of the cylinder. If the triangle does not contain the inside surface, then it can be deformable to a point; that is, the closed path formed by the sides of the triangle is a loop that can be contracted to a point, as can all the triangles in the plane. Thus, triangles that cannot contain the cone point or the tube of the cylinder are identical to plane triangles, and so SAS holds for those triangles.

Definition 6

A triangle on a sphere is small for SAS if it has distinct vertices, no angles of 180°, no sides longer than a great circle, and of the possible geodesic segments between the two unconnected vertices, the undetermined side of the triangle is the closer one that is opposite to the Angle of Side-Angle-Side (in the sense that it is the first segment reached by a ray emanating from the Angle of Side-Angle-Side).

Proof:

This definition includes a class of triangles that do not look like "normal" triangles. For instance, these triangles can intersect themselves, and we can have what some students call "fish" triangles:

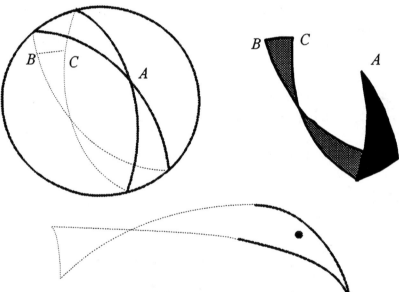

The requirement of no 180° angles makes the triangle small enough for SAS even though the sides can be large enough to allow the triangles to intersect themselves. The choice of the closest side opposite the Angle of Side-Angle-Side ensures that the side that connects the two points is uniquely determined.

Definition 7

A triangle on the cone or cylinder is small if it can be "unrolled onto the plane" (or lifted to a covering space).

Proof:

Such triangles **are** plane triangles and thus have all the properties of plane triangles.

Some students will devise properties of triangles on the cone and cylinder that will imply that the triangles can be unrolled onto the plane (or lifted to a covering space). For example:

Definition 8

A triangle on a cylinder is small if it does not encircle the cylinder or if it has finite area. A triangle on a cone is small if it does not encircle the cone or if it has finite area and does not contain the cone point.

In addition the students may envision other classes of triangles which are not small, but which nevertheless satisfy SAS. For example:

All triangles with precisely one side longer than 1/2 a great circle satisfy SAS.

Students may arrive at more examples than those we mentioned above, and the instructor should be open to all of the different definitions that may appear. Students are not required to create the best definitions ("best" in the sense of getting the strongest theorem possible), but only to find a class of triangles on the surfaces that will make the theorems true.

As you may have already noticed in Problem 5, students are generally resistant to using covering spaces. This will most likely continue to be the case in Problem 6. Yet, covering spaces are a helpful tool for thinking intrinsically. The instructor can challenge students to think intrinsically by illustrating how some triangles, even though they look strange extrinsically, will, for the bug, look intrinsically like reasonable triangles. One way to do this is to ask the students to imagine that they are on a walk in the forest. How will they determine in the course of the walk if they have walked in a triangular path? In Figure 5.11 on page 56 of the main text we give an example of an extrinsically strange triangle that is a normal triangle intrinsically.

The instructor should be aware that students are unaccustomed to making their own definitions. Usually the students are asked to think about the mathematics *given the definitions.* At first they may become confused and somehow feel incapable of thinking so freely. However, once they begin creating their own definitions they will start to enjoy their new-found freedom.

Example of Students' Work on Problem 6

Professor and T.A. comments: Student answer:

Problem 6 - Semyon Kruglyak

The reason that SAS holds on the plane is that there is only one way to connect two points with a straight line. Therefore, if one has the following:

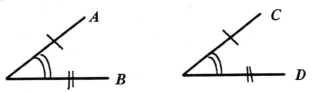

A, B, and *C, D* could only be connected by the same line and the triangles would be congruent. On a sphere, this does not hold. *A, B* could be connected as shown, while *C, D* could be connected by

Good

using a straight path back around the sphere. The sum of the angles of
the latter triangle would exceed 180°, but that is possible on a sphere.
On a cylinder the same is true. If we imagine that this paper is an
unrolled cylinder the following are two possible triangles:

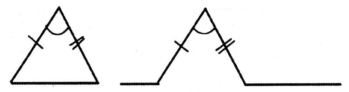

*Is SAS true for
small triangles?*

The same reasoning applies to a cone. SAS does not hold on these
figures because there is more than one way to connect two points with
a straight path.

Response to comments:
   SAS will hold on every surface if the shortest distance is always
picked to connect the two points. (No going around the back.) Direct
congruency will only occur if the triangles are not mirror images of
each other. If they are:

Why?

What if:

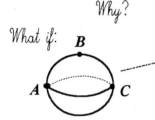

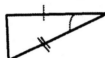 

then flipping or reflecting is required and we have congruence, but not
direct congruence.

Response to comments:
   A triangle is a closed figure consisting of three points connected by
three geodesics.
   In its original form, SAS only holds on the plane (as explained
earlier). We can, however, define a small triangle such that SAS will
hold on every surface.

Nice

   Sphere: <u>A triangle is small if it is not intersected by every great
circle.</u> Any triangle that goes all the way around the sphere is,
therefore, eliminated.
   e.g.

intersects every great circle

   There are usually two ways to connect two points on a sphere with a
geodesic. This restriction takes away one of them and leaves a triangle
to be uniquely determined by SAS. The case where the two points are
poles with an infinite number of connections is also eliminated because
all great circles would intersect a side of such a triangle.

On a cone and cylinder, a triangle is small if it is not intersected by
every generator. The same reasoning applies as used with the sphere.
Any looping, going around back, or spiraling is eliminated. Other
characteristics of a small triangle include: angles must be on the inside,
area must be finite, and the cone point must not lie within.

Me too!   However, I think that small (triangles) meaning not intersected by
every great circle or generator works really well.

PROBLEM 7. *Angle-Side-Angle (ASA)*

*Are two triangles congruent if one side and the adjacent angles of one are congruent to
one side and the adjacent angles of another?*

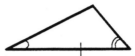

Let us look at a proof of ASA on the plane: Pick one of the vertices of the Side of Angle-
Side-Angle on one of the triangles, and translate that vertex to the corresponding vertex of the
other triangle. Next, rotate the given side until the congruent sides on the two triangles coincide.
Now either the congruent adjacent angles are already coinciding, or we must reflect one triangle
using the given side as an axis. Because both pairs of angles are given as congruent they will now
coincide. The question here is: *Do the undetermined vertices (A and A ' in the figure below)
coincide?* If the triangles lie on the plane, then they will coincide because *on the plane two different geodesics intersect in at most one point.*

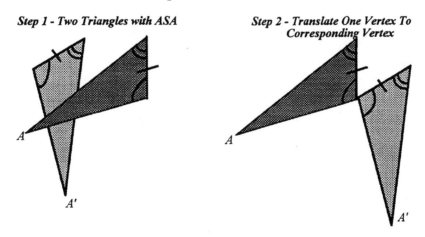

Step 1 - Two Triangles with ASA

*Step 2 - Translate One Vertex To
Corresponding Vertex*

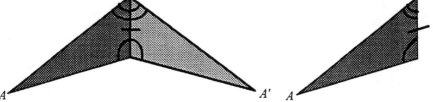

Step 3 - Rotate Until Congruent Sides Correspond   Step 4 - Reflect about Congruent Side

The planar argument for ASA does not work on spheres, cylinders and cones because, in general, geodesics on these surfaces intersect in more than one point. As was the case for SAS, the question now is, can we find a class of small triangles on each of the different surfaces for which the above argument is valid? Earlier in this chapter we illustrated several classes of triangles for which SAS is true. Students should check if their previous definitions of a small triangle are too weak, too strong, or just right to make ASA true on spheres, cylinders, and cones. The instructor can suggest that they look at cases for which ASA does not hold. Just as with SAS, some interesting counter-examples arise. Let us look at some of these counter-examples:

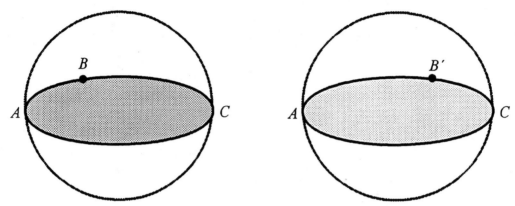

The counterexample in this figure shows two triangles, $\triangle ABC$ and $\triangle AB'C$. Although they have one side and the corresponding adjacent angles congruent, the two triangles are not congruent. Euclid did not accept as angles what we call "180° angles." He did not refer to "a straight angle," but instead would say "two right angles." Also keep in mind that it is natural if students do not see the triangle in the above figure as a triangle, because on the plane the sides of a triangle always lie on different geodesics; and so form this point of view the example does not look like a triangle.

Let us look at another counterexample for ASA:

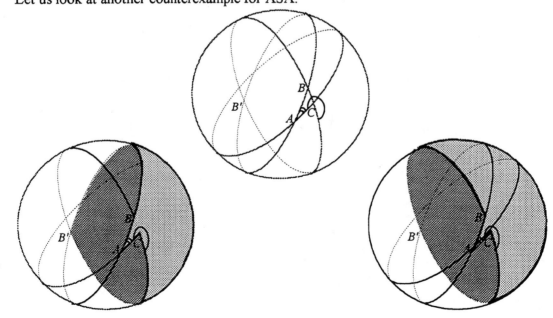

In the figure above both triangles have one side and the adjacent angles congruent, but the triangles are not congruent. So, in general, ASA is not true for large triangles. Do any of the definitions of small triangles that were used for SAS work for ASA? At the very least, students will have to require that *two consecutive sides cannot be collinear.*

For triangles in an open hemisphere or in an open 1-sheeted covering of a cone or cylinder ASA is true. Proof: Through reflections, translations, and rotations we can make the side and adjacent angles of ASA coincide. Now we must prove that the vertices coincide. The two triangles are contained in the same open hemisphere, and inside of any open hemisphere two great circles intercept only once. On a cylinder or cone the triangles are planar triangles and consequently they will satisfy ASA.

In the following problems, students will be proving more properties of triangles on the plane and will explore the possibilities of expansion of those properties to spheres, cones and cylinders.

Example of Student's Work on Problem 7

Professor and T.A. comments:

Student answer:

Problem 7- Christos Ioannou

This theorem does not hold on any surface:
$\angle CAB \cong \angle C'A'B'$, $AB \cong A'B'$ and $\angle ABC \cong \angle A'B'C'$ then
$\triangle ABC \cong \triangle A'B'C'$

ABC has to be an isosceles triangle since $\angle ABC \cong \angle CAB$
$(\cong \angle A'B'C')$

Which ASA are you indicating? Angle A is not congruent to Angle A'

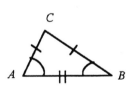

Obviously $\triangle ABC$ and $\triangle A'B'C'$ are not congruent even though they hold the three properties stated above

Response to comments:

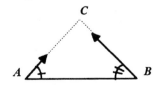

What we really have is an angle-side-angle construction
This is fixed and we would like to see if the place where the two rays will meet is specified.

On the plane, the two rays have to converge; that is, angle at vertex A and angle at vertex B have to be such that the sum is less than 180°.

Good

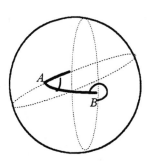

Given our definitions in Problem 6, the restriction of $\angle A + \angle B < 180°$ has to hold for the sphere also because the two rays will meet at two points: one of which is closest to A rather than going the other way, and the other which is closest to B.

So this cannot be defined as a triangle, it can only be defined as one if $\angle A + \angle B < 180°$.

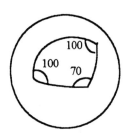

Thus ASA holds for $\angle A + \angle B < 180°$, since this is the only way in which a triangle can be defined. Actually some triangles on a sphere can be defined for angles $> 180°$:

Since we cannot say that ASA holds for any triangle we say it does not hold. It does hold, though on a sphere for triangles which are defined on 1/2 of the sphere.

For cylinders and cones ASA does not always hold, since a triangle is not always defined by an ASA construction.

For cone $>$ than 360° we may have two points which are unjoinable, which goes against our definition of triangle.

For small triangles on cone and cylinder?

Response to comments:

For small triangles on the cylinder or cone, if the whole triangle can be seen we have an ASA construct.

For the triangle to be small $\angle A$ and $\angle B$ must have $\angle A + \angle B < 180°$ or else C will not be visible.

This means that we can simply unroll the shape and the triangle will be on a plane with all the angles remaining the same and all the lines remaining straight. Since ASA holds on the plane, it holds for small triangles on cones and cylinders because these small triangles have the same properties as on the plane.

Chapter 6

Area, Parallel Transport, and Holonomy

This chapter includes Problems 8 to 11, which begins with the area of a triangle, then introduces students to the notion of holonomy and culminates in a discussion of the Gauss-Bonnet formula. The result of Problem 11 is a part of dissection theory and is necessary for Problem 10. Dissection theory will be explored further in Chapters 12 and 13.

PROBLEM 8. *Area of a Triangle on a Sphere*

Show that on a sphere the formula Area $(\Delta) = [\ \Sigma \angle's - \pi\ ]\ A/4\pi$ *makes sense, where A is the area of the sphere and* $\Sigma \angle's$ *is the sum of the angles of the* $\Delta\square$ *in radians.*

The expression $\Sigma\angle's - \pi$ is called the *excess* of the triangle. The following hints are given to the students: (1) Prove that the area of a lune with angle α is $2\alpha R^2$, and (2) Note that the sides of a triangle on a sphere divide the sphere into overlapping lunes.

Proof:

Consider ΔABC, a triangle that does not self-intersect and that has no collinear vertices. Let A', B' *and* C' *be* the points opposite A, B and C, respectively. Then we know that $\Delta ABC \cong \Delta A'B'C'$. We can see that the corresponding angles are congruent by using the Vertical Angle Theorem twice, and by the fact that a lune has congruent angles. Since the triangles satisfy the conditions of AAA, they are congruent. Furthermore, given that ΔABC can be seen as the intersection of three lunes namely, L_α, L_β and L_γ, $\Delta A'B'C'$ can also be seen as the intersection of the opposite lunes, L_α', L_β' and L_γ'. Note that the sphere is covered by these six lunes, which are disjoint except for the fact that both ΔABC and $\Delta A'B'C'$ are each covered three times. Now, if we let $A(\Delta)$ denote the area of the congruent triangles, then the whole area of the sphere can be expressed in the following way:

$$A(S) = [A(L_\alpha)+A(L_\beta)+A(L_\gamma)-2A(\Delta)] + [A(L_\alpha')+A(L_\beta')+A(L_\gamma')-2A(\Delta)].$$

Since the area of a lune is proportional to the area of the whole sphere in direct relation to its angle, we know:

$$A(L_\theta) = (\theta/2\pi)\, A(S).$$

Then, combining these equations, we get the following intrinsic expression:

$$A(\Delta) = [\Sigma\angle's - \pi]\, (A(S)/4\pi).$$

Finally, since $A(S) = 4\pi R^2$ (see Problem 8), we know that the area of the triangle is given by:

$$A(\Delta) = R^2\, [\Sigma\angle's - \pi].$$

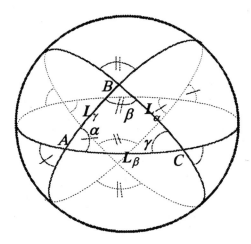

Example of Students' Work on Problem 8

Student answer:

Problem 8 - C. Earle Youngdahl

(On the sphere) Area(Δ) = $[(\sum \angle$'s$)-\pi]R^2$, where R is the radius of the sphere and $(\sum \angle$'s$)$ is the sum of the angles of the Δ in radians. (Area of the whole sphere is $4\pi R^2$.)

Since we know the area of the whole sphere is $4\pi R^2$, we can assume that this is a lune with angle 2π. We can know that the area of a lune must be proportional to the area of the sphere. Given a lune of angle α, we have:

$$\{A(lune) = area\ of\ lune\}$$
$$A(lune)\ /\ A(sphere) = \alpha/2\pi$$
$$A(lune)\ /\ 4\pi R^2 = \alpha/2\pi \implies A(lune) = 4\pi R^2/2\pi$$
$$A(lune) = 2R^2\alpha$$

To get the area of a triangle we set up three lunes that share a common triangle:

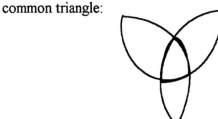

(It's kind of hard to draw accurately.)

The three biangles form 2 antipodal triangles that are congruent by AAA. The symmetric properties of the sphere leads me to believe that each figure formed by the three lunes is equal in area to 1/2 area of the sphere.

We can state:

$$1/2\ A(sphere) = A(3\ lunes) - 2A(\Delta).$$

{We must subtract twice the area of the △ because it is counted once for each lune.} {assuming lunes w/ angles α, β, γ respectively}
Substituting:

$$1/2 \, (4\pi R^2) = 2R^2 \, \alpha + 2R^2 \, \beta + 2R^2 \, \gamma - 2A(\Delta)$$
$$2\pi R^2 - 2R^2 \, (\alpha+\beta+\gamma) = -2A(\Delta)$$
$$R^2(\alpha+\beta+\gamma-\pi) = A(\Delta)$$

Q.E.D.

Introducing Parallel Transport

One way to introduce parallel transport is to ask students to imagine that they are walking along a straight line or geodesic carrying a stick that makes an angle with the line they are walking on. If they walk along the line, maintaining the direction of the stick relative to the line constant, then they are doing a ***parallel transport*** of that "direction" along the path.

The notion of holonomy is one that most students seem drawn to. Experimenting with calculating the holonomy of different triangles on the sphere helps students to understand parallel transport.

PROBLEM 9. *The Holonomy of a Small Triangle*

In Problem 9, the students are introduced to the notion of the holonomy of a triangle.

DEFINITION: *The **holonomy of a small triangle**, H (△), is defined as follows:*
If you parallel transport a vector (a directed geodesic segment) counterclockwise around the three sides of a small triangle, then the holonomy of the triangle is the smallest angle measured counterclockwise from the original position of the vector to its final position.

The construction of the holonomy of △*ABC*, that is, the construction of the angle H. (△), is represented in the figure below:

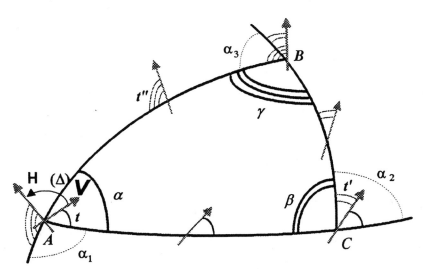

Show that the following is true on a sphere: The holonomy of a small triangle is equal to 2π *minus the sum of the exterior angles and is the same as its excess.*

Let α, β and γ be the interior angles of the triangle and α_1, α_2, α_3, the exterior angles, then algebraically the above statement can be written as:

$$H\ (\Delta) = 2\pi - (\alpha_1 + \alpha_2 + \alpha_3) = (\alpha + \beta + \gamma) - \pi$$

Proof: Let us choose A as a starting vertex. At vertex A we have that:

$$H\ (\Delta) + t + \alpha_1 + t'' = 2\pi.$$

On the other hand, given the way the vector **V** is parallel transported along the triangle, we can conclude that:

$$t' = \alpha_2 - t, \ \text{ and } \ t'' = \alpha_3 + t', \ \text{ and so } \ t'' = \alpha_3 + \alpha_2 - t.$$

Combining these equations we get:

$$H\ (\Delta) = 2\pi - (\alpha_1 + \alpha_2 + \alpha_3).$$

Then, by definition:

$$\alpha_1 = \pi - \alpha, \ \ \alpha_2 = \pi - \beta, \ \ \alpha_2 = \pi - \gamma$$

and finally we have:

$$H\ (\Delta) = (\alpha + \beta + \gamma) - \pi = \Sigma\angle's - \pi$$

Note that one consequence of this proof is that the holonomy does not depend on either the vector or vertex we start with. This is to be expected, since parallel transport does not change the relative angles of any figure. Using this fact, we can perhaps get a clearer picture of the holonomy by deciding that our initial vector will lay along one of the sides:

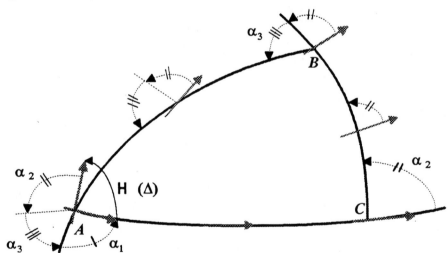

Example of Students' Work on Problem 9

Student answer:

Problem 9 - Ryan Gallivan

To prove that $H\ (\Delta)R^2 = R^2[2\pi - (\alpha_1 + \alpha_2 + \alpha_3)] = \text{Area}(\Delta)$, we need to prove two equalities:

(1) $R^2\ [2\pi - (\alpha_1 + \alpha_2 + \alpha_3)] = \text{Area}(\Delta)$

(2) $H\ (\Delta) = [2\pi - (\alpha_1 + \alpha_2 + \alpha_3)]$

(1) $R^2[2\pi-(\alpha_1+\alpha_2+\alpha_3)] = R^2[2\pi - (\pi-\alpha_1'+\pi-\alpha_2'+\pi-\alpha_3')]$
where α_i' is the interior angle adjacent to α_i

$$= R^2[2\pi - (3\pi - (\alpha_1' + \alpha_2' + \alpha_3'))]$$
$$= R^2[2\pi - 3\pi + \alpha_1' + \alpha_2' + \alpha_3'))]$$
$$= R^2[-\pi + \alpha_1' + \alpha_2' + \alpha_3'))]$$
$$= [\textstyle\sum \angle\text{'s} - \pi]\, R^2$$
$$= \text{Area}\,(\Delta) \qquad \text{(from 8, even though I haven't proved it yet)}$$

(2)

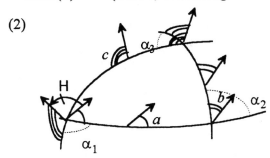

α_2 is the sum of two angles: a and b. Angle c is the sum of α_3 and b.

 $\qquad\qquad X + Y = a + b = \alpha_2$

 This vector sweeps out 2π radians (also the number of radians in a circle)

So $H(\Delta)$ is $2\pi - (\alpha_1 + \alpha_2 + \alpha_3)$ and $H(\Delta)R^2 = R^2[2\pi - (\alpha_1 + \alpha_2 + \alpha_3)]$

The above discussion of holonomy is in the context of small triangles on a sphere, but the results have a much more general applicability that is a major aspect of differential geometry. In particular, we can write the result from Problem 9 in the following form:

$$H(\Delta) = 2\pi - (\alpha_1 + \alpha_2 + \alpha_3) = A(\Delta)\, R^{-2},$$

which traditionally is called the **Gauss-Bonnet Formula**. The quantity R^{-2} is traditionally called the **Gaussian curvature** or just plain **curvature**.

PROBLEM 10. *The Gauss-Bonnet Formula for Polygons on a Sphere*

The Gauss-Bonnet Formula not only holds for small triangles, but can be extended to any small (i.e., contained in an open hemisphere), simple (i.e., non-intersecting) polygon (i.e., a closed curve made up of a finite number of geodesic segments) contained on a sphere.

DEFINITION. *The **holonomy of a small simple polygon**, H (Γ), is defined as follows:*
If you parallel transport a vector (a directed geodesic segment) counterclockwise around the sides of a small simple polygon, then the holonomy of the polygon is the smallest angle measured counterclockwise from the original position of the vector and its final position.

If you walk around a polygon with the interior of the polygon on the left, the exterior angle at a vertex is the change in the direction at that vertex. This change is positive if you turn counterclockwise and negative if you turn clockwise.

Show that if Γ *is a small simple polygon on a sphere, then*

$$H\ (\Gamma) = 2\pi - \Sigma\ \alpha_i\ = A(\Gamma)\ 4\pi/A = A(\Gamma)\ R^{-2},$$

where $\Sigma\alpha_i$ *is the sum of the exterior angles of the polygon.*

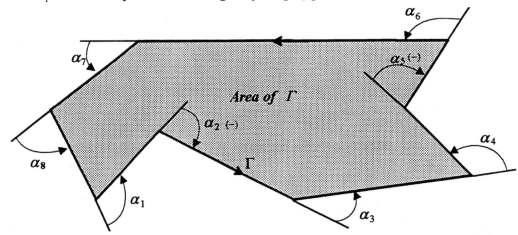

Outline of a proof: Divide the polygon into small triangles. It is possible to do this by constructing geodesic segments in the interior of the polygon without adding any new vertices (see Problem 11). Show, by constructing geodesic segments in the interior one at a time, that the holonomy of the polygon is the sum of the holonomies of the triangles. In addition, check directly (as was done at the end of the proof of Problem 9) that $H\ (\Gamma) = 2\pi - \Sigma\ \alpha_i$.

Example of Students' Work on Problem 10

Professor and T.A. comments: Student answer:

Donna Muise - Problem 10
Prove the Gauss-Bonnet Formula for polygons on a sphere:
GB: $H\ (\Gamma) = 2\pi - \Sigma\alpha_i = A(\Gamma)R^{-2}$

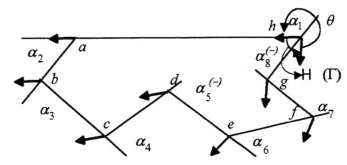

At a, α_2
At b, $\alpha_2 + \alpha_3$
At c, $\alpha_2 + \alpha_3 + \alpha_4$
etc....
At h, $\alpha_2 + \alpha_3 + \alpha_4 - \alpha_5 + \alpha_6 + \alpha_7 - \alpha_8 + \alpha_1 = $ angle θ
angle $\theta = \Sigma\alpha_i \Rightarrow H\ (\Gamma) = 2\pi - \Sigma\alpha_i$

2nd part:

$H\ (\Gamma) = A(\Gamma)R^{-2}?$

Response to comments:
Some random polygon:

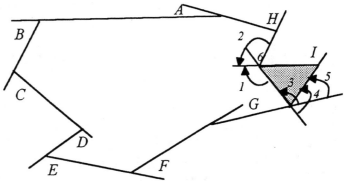

Assume: $H\ (\Gamma) = 2\pi - \Sigma\alpha_i = A(\Gamma)R^{-2}$ for the non-shaded polygon with n vertices. Now prove for the original polygon plus the shaded polygon with $n+1$ vertices, it works that $H\ (\Gamma) = 2\pi - \Sigma\alpha_i = A(\Gamma)R^{-2}$ (again, for the new polygon, the shaded plus the non-shaded). We know that for the shaded polygon $H\ (\Gamma) = 2\pi - \Sigma\alpha_i = A(\Gamma)R^{-2}$ and we also know that:

(Area of the shaded polygon) + (Area of the non-shaded polygon)
= Area of the entire figure.

Area of shaded polygon = $(2\pi - \Sigma\alpha_i)R^2$
Area of non-shaded polygon = $(2\pi - \Sigma\alpha_i)R^2$ (as assumed)

$\Sigma\alpha_i$ for shaded polygon = $1 + 4 + I$
$\Sigma\alpha_i$ for non-shaded polygon = $A + B + C - D + E + F - G + H + 3 - 6$
$\Sigma\alpha_i$ for entire figure = $A + B + C - D + E + F - G + H + I + 5 - 2$

From polygon: $2 - 6 + 1 = \pi$
$$4 + 3 - 5 = \pi$$

Area of non-shaded + Area of shaded $= (\pi + 5) + (\pi - 2)$
$R^2(4\pi - (1 + 4 + I + A + B + C - D + E + F - G + H + 3 - 6))$

$1 - 6 = \pi - 2$
$4 + 3 = \pi + 5$

Area of non-shaded + Area of shaded =
$R^2(2\pi - (A + B + C - D + E + F - G + H + I + 5 - 2))$
$= (2\pi - \Sigma\alpha_i)R^2 = A(\Gamma)$

Since this works for triangles and I can prove $n + 1$ from n, this will work for any polygon with any n vertices. In effect, if I didn't assume a polygon with n vertices, I could start with the triangle only as my polygon and add a triangle to that (thereby adding one vertex) and I would know it works for both triangles. Prove it works with polygons with 4 vertices and then add another triangle to that, etc., to prove that $A(\Gamma) = (2\pi - \Sigma\alpha_i)R^2$ for any polygon.

PROBLEM *11. Dissection of Polygons into Triangles*

Prove that every simple polygon can be dissected into small triangles without adding extra vertices.

There are in the literature many incorrect descriptions of dissections of polygons, several by well-known mathematicians. For a discussion of these errors see the article: Chung-Wu Ho, Decomposition of a Polygon into Triangles, *Mathematical Gazette*, vol. 60 (1976), 132-134. You may want to point this out to the students at an appropriate time.

Dissecting convex polygons: A polygon is **convex** if any two points on it can be joined by a geodesic segment lying entirely in its interior. To dissect a convex polygon, pick any vertex and join it to all the others that are not adjacent. This will dissect the convex polygon into triangles. On a sphere if these triangles are not small then dissect them further.

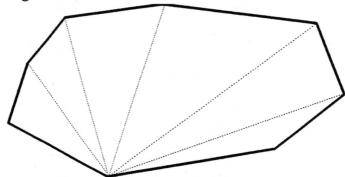

Dissecting concave vertices: If the polygon is not convex, then it has at least one concave vertex. We say that a vertex is **concave** if its (interior) angle is greater than a straight angle. To dissect a polygon that has concave vertices, at each concave vertex cut along a segment which divides the interior angle into two convex (less than or equal to a straight angle) angles. Now the polygon will be dissected into a finite number of convex polygons, and then each of the convex polygons can be dissected into (small) triangles.

Parallel cuts: Pick any line (great circle) *l*. For each vertex of the polygon draw a line (great circle) through the vertex and perpendicular to *l*. If you now cut along the intersection of these lines with the interior of the polygon, the polygon will be dissected into triangles and quadrilaterals, each of which can be dissected into two triangles.

Dissecting into triangles without adding any new vertices: On a sphere, if there are no convex vertices then the exterior of the polygon is convex. In this case, pick a point *p* that is the opposite pole of some point in the exterior of the polygon and cut along all the short great circle segments joining *p* to the vertices of the polygon.

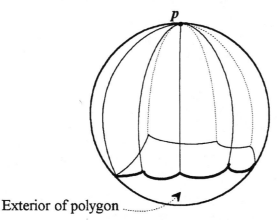

Exterior of polygon

If there is a convex vertex *v*, then let *l* be a geodesic joining the two adjacent vertices. In the diagram below, *v′* and *v″* are adjacent to *v*. If the segment of *l* between the two vertices lies totally inside the polygon, then make a cut along it. If not, then parallel transport *l* toward *v* along *vv′* until the portion of *l* within Δ*vv′v″* intersects only interior portions of the area of the polygon except for vertices, such as *w*, which lies on the polygon's perimeter. Now, *l* must contain a vertex *w* of the polygon (lying in the interior of Δ*vv′v″*). Cut along the segment *vw*.

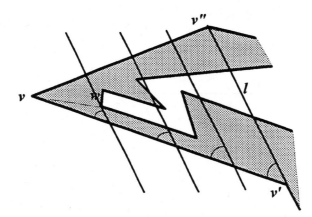

In both cases, we have cut the polygon into two polygons each with fewer vertices. Continue until only triangles are left.

You can check that this method dissects the polygon into $n-2$ triangles where n is the number of sides of the polygon. Then, one can use Problem 20 to show that:

On the plane, the sum of the angles of a polygon with n sides is $(n-2)\pi$.

Example of Students' Work on Problem 11

Professor and T.A. comments:

Student answer:

Joanne Galinsky - Problem 11

Pick a corner. Draw lines from it to other corners until you have exhausted all possibilities. Check if you have successfully divided the figure into triangles. If not, pick another corner and repeat procedure until you are done.

Sometimes, you will end up with a quadrilateral or some other figure in the interior. If this happens, use the same procedure on this smaller polygon.

How can you be sure that

— you are making triangles?

— the triangles are inside the polygon? No

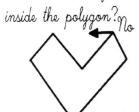

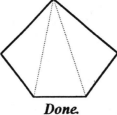

Done.

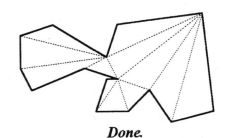

Done.

How are you sure that, by cutting selected corners, you have a convex polygon?

Response to comments:

I originally asserted that by picking a corner and connecting it to all other corners, and then continuing, we could dissect any polygon into triangles. I need to refine this a bit, to ensure that I am making triangles inside the polygon.

Here's the new method:

1). Chop the polygon into simpler, convex (Is that the right word?) polygons by connecting select corners. By splitting the figure into many simple polygons, you ensure that when you connect corners, the resulting triangles will be inside the polygon.

2). Once you have split the polygon into simple polygons, you can pick a corner of a polygon, connect it to as many others as possible, and then move to another corner, until within this simple polygon you have all triangles. Then move to the next polygon and repeat.

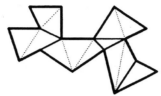

The polygon shown above (dark lines) has corners that, connected to others, will result in triangles outside the polygon. To resolve this, first cut the shape into smaller, simple polygons (lighter, solid lines). Then, connect corners of each smaller polygon (dashed line).

Response to comments:

I need to state more specifically how to connect corners to separate a figure into a number of simple convex polygons:

(1). If you see a "concave corner," connect it (with a line that stays in the interior of the figure) to another corner. Check to see if the region you have partitioned off is convex. If it isn't, connect the same corner to a new one until the region is convex. Then, move onto the next "concave corner" and repeat the procedure.

Is this always possible? Why? Is it necessary?

(2). Once you have split the polygon into convex polygons, you can pick a corner of a polygon, connect it to as many others as possible, and then move to another corner, until within this polygon you have all triangles. Then move to the next polygon and repeat.

Will this always happen?

Response to comments:

In class we discussed various ways to cut up a polygon into a number of simple, convex polygons. I liked Karen's suggestion of following the perimeter of the polygon and "going straight" whenever you come to a concave angle. This method ensures that the remaining figure will consist only of convex polygons, because you have gotten rid of every concave angle. Once you have split the figure into convex shapes, then you can split each small polygon into triangles by picking a corner, connecting it to as many others as possible, and then picking another corner until you have all triangles.

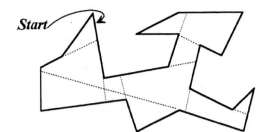

Chapter 7

ITT, SSS and ASS

PROBLEM 12. The Isosceles Triangle Theorem (ITT)

Given a triangle with two of its sides congruent, then are the two angles not included between those sides also congruent?

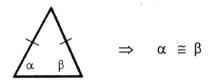

A standard proof of ITT for all triangles on the plane is the following: Bisect the angle formed by the congruent sides of ITT. Then we have two triangles which have one common side (the angle bisector), the congruent angles of the angle bisection, and the congruent sides given by ITT. So, we can use SAS which we now know to be true for all triangles on the plane. Since the triangles are congruent, the other two angles of the original triangle must also be congruent.

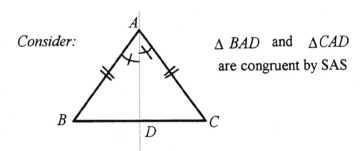

Consider: Δ *BAD* and Δ*CAD* are congruent by SAS

For the other surfaces, it suffices to use the definitions of small triangle that we used for SAS. Then the planar proof can be extended to prove:

ITT is true for small triangles on a sphere, cylinder, or cone.

Note that the planar proof of ITT leads to a corollary:

The bisector of the top angle of an isosceles triangle is also the perpendicular bisector of the base of that triangle.

However, there are counter-examples for ITT for large triangles on cones and cylinders. Let's look at an example:

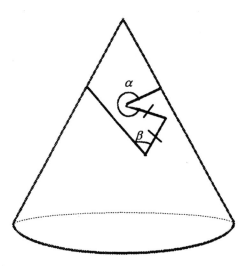

An isosceles triangle on a cone

The instructor should encourage students to try to find counter-examples for ITT as they try to prove it for each of the surfaces. For example, students will most likely prove ITT on a sphere using a definition of small triangles and SAS as we suggested above, even though ITT is true for *every* triangle on a sphere. In this case, the instructor should encourage students to enlarge the class of triangles they have chosen for their proof or to find a counterexample for large triangles.

Let us now look at a symmetry proof of:

*ITT is true for **all** triangles on a sphere.*

Proof: Consider an isosceles triangle, $\triangle ABC$, with sides $AB \cong CB$ (see the figure below). Bisect $\angle ABC$ and let D be the point of intersection of this angle bisector with the base AC of the triangle. Reflect the triangle about the angle bisector, so that the congruent sides of the triangle exchange places. The vertices A and C together with the point D define a unique great circle. Note that, with the reflection, the point B does not move from its place but the vertices A and C exchange places. Thus, the great circle, uniquely defined by those three points, A, C, and D, goes onto itself. Therefore, the angles adjacent to the congruent sides, $\angle CAB$ and $\angle ACB$, are congruent.

This proof is also valid on the plane and, indeed, on any surface for which a reflection through the angle bisector is a global symmetry of the surface.

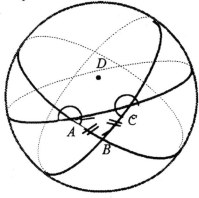

At first, students may have difficulty coming up with this symmetry proof of ITT on their own. As we said before, most students will use SAS when trying to prove ITT on the plane. Since they gave a lot of thought to SAS in Problem 6, it is natural that they choose the same definition of small triangle on a sphere here. They might not try to look for counterexamples involving large triangles, so you should suggest that they look for such counterexamples.

Now let's look at some other proofs of ITT. For those who are uncomfortable with the symmetry proof above, the proofs below are students' proofs. Students do not usually come up with a symmetry proof for this result:

If ITT is true for small triangles on a sphere then it is true for all non-self-intersecting triangles on a sphere.

Proof: If *AB* and *CB* are congruent geodesic segments of length less than half of a great circle, then there are two possible choices of segment for the third side and two possible choices for the top angle ∠*ABC* (see the figure below). Thus, there are four choices of isosceles triangles with sides *AB* ≅ *CB*. If γ and δ are the congruent base angles of the smallest of these isosceles triangles, then their supplements α and β are also congruent. The angles φ and η, opposite to α + γ and β + δ and both 180°, are also congruent. So the pairs of base angles of the four isosceles triangles are γ, δ ; α, β ; γ + φ, δ + η ; and α + φ, β + η. All of these base angle pairs are congruent.

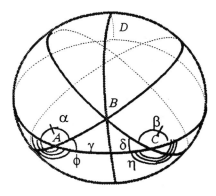

Another proof:

Consider Δ*ABC* and the triangle obtained by reflecting Δ*ABC* (see the "mirror triangles" in the figure below). These two triangles satisfy the conditions of SAS and consequently Δ*ABC* ≅ Δ*ACB*. But then, ∠*ABC* ≅ ∠*ACB*.

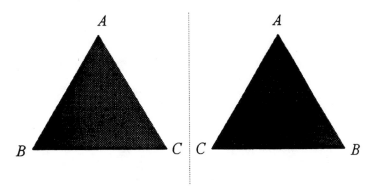

Examples of Students' Work on Problem 12

Professor and T.A.   Student answer:
Comments:

Problem 12 - Anonymous[†]

ITT: If two sides of a triangle are the same, then the two angles they make with the third side must be the same.

Restatement: Two straight rays of equal length emanating from the same point will meet the straight line between their endpoints at equal angles.

Proof in the plane: Draw the line bisecting the angle included by the two equal sides, and extend it to the third side. The two triangles thus formed have SAS, and thus are congruent.

Good!

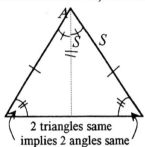

2 triangles same
implies 2 angles same

This proof depends on the fact that the bisector of the included angle will meet the third side of the triangle, and the fact that SAS is true in the plane. The proof can clearly be extended to "small" triangle on the cone or cylinder, as we have shown that these are always triangles in the plane as well.

Is ITT false for any triangle on the sphere?

On a sphere, the bisector of the included angle will meet the third side of the triangle twice, as all straight lines intersect twice on the sphere. One of the paths from the included angle to the third side will be within the triangle and one will be outside it, as those are the only available options. The path within the triangle can be used in to demonstrate SAS on the sphere, and thus to show ITT is true.

Put this on an actual sphere

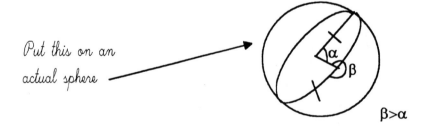

β>α

Response to comments:

ITT for "large" triangles on the sphere of the type

(where the angle between the 2 equal sides is > 180°)

[†]Name withheld at the request of the student.

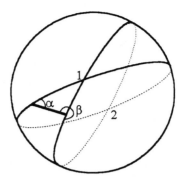

In the above picture both triangles can be formed, given the common side and α and β. In both cases, and in the two other cases of pairs of large triangles drawn, one side of any triangle (other than the common side) must go to the first intersection point (of the two great circles made by extending rays at α and β from the common segment, until they become great circles) and the other side goes through two intersection points. This is necessary for the two sides to meet. If each side went from its start to the first intersection point, then the two sides would have different intersection points, separated by 1/2 great circle, or 180° as we know from Problem 6.

Since 1 side goes through only one intersection point, it must be of length < 1/2 great circle. The side going through two intersection points must be bigger or equal than 1/2 great circle. Thus, two sides of such a triangle cannot have the same length. The only conceivable way for this to occur is if one of the intersection points is at the point of origin of the two rays, that is, if side A has length 0. In this case, each ray would start at one intersection point, and end at the other, making them both 1/2 great circle in length. However, this would create a two-sided figure, not a triangle.

Thus, ITT is not contradicted by the case of triangles with 1 angle > 180°, because triangles with two equal sides cannot be constructed if 1 included angle is < 180° and the other is > 180°.

Both sides are equal

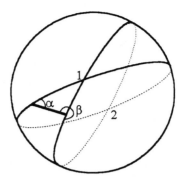

and obeys ITT

ITT holds for small triangles because each side of a small triangle is directly opposite (as in Sylvan's definition) one of its angles. This means each angle can be bisected by a ray which will meet the corresponding side of the triangle (of the [figure] type).

Can you make a proof for ITT *for all large triangles?*

ITT also holds for large triangles because large triangles cannot be constructed such that two sides are equal except in the case of one of the angles being 180°.

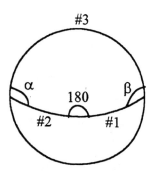

In this case the sides #1 and #2 are collinear, so the angles α & β are = iff two great circles intersect in equal angles; which they do, by the VAT on a sphere.

Thus, ITT holds for this case.

I believe, therefore, that ITT holds for all triangles on the sphere.

Addendum: Large triangles not of the type 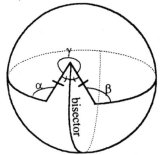 can bisect the angle included between the equal sides, which implies the same proof as for small triangles, since SAS holds for large triangles too.

What about this angle? Can you form a triangle with α, γ, β?

Can bisect the angle included between the 2 equal sides ⇒ Same proof as for small triangles, since SAS holds for large triangles, too.

Thus, ITT holds for large triangles with the included angles is < 180°. And, ITT holds for small triangles and large triangles with included angle ≤ 180°.

Response to comments:
Large triangles on sphere:
Case A:

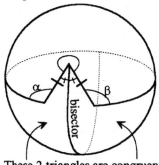

These 2 triangles are congruent by SAS

Case B: These two triangles are congruent

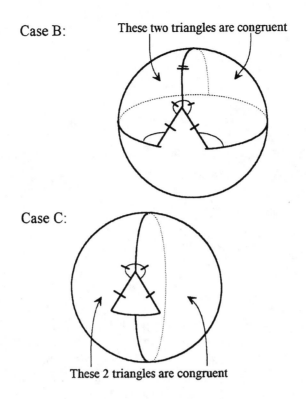

Case C:

These 2 triangles are congruent

Therefore, ITT is true for all large triangles on the sphere.

PROBLEM 13. *Side-Side-Side (SSS)*

Are two triangles congruent if the two triangles have congruent corresponding sides?

Depending on how one chooses to prove it, SSS relies on either ITT or on corollaries of ITT and symmetry of circles. We will look at two proofs of SSS, both of which apply to the plane and sphere.

1st proof:

Consider $\triangle ABC$ and $\triangle A'B'C'$ such that $AB \cong A'B'$, $AC \cong A'C'$ and $BC \cong B'C'$. Using translation, rotation, and (if necessary) reflection, make two corresponding sides of the triangles coincide as in the figure below. Consider the geodesic that joins the vertices B and B' and the isosceles triangles $\triangle BAB'$ and $\triangle BCB'$. We have that $\angle B'BC \cong \angle BB'C$ and $\angle ABB' \cong \angle AB'B$. Now, if $\angle ACB < 90°$, then $\angle B'BC$ and $\angle ABB'$ are part of and complete $\angle ABC$, and $\angle BB'C$ and $\angle AB'B$ are part of and complete $\angle A'B'C'$. Consequently, $\angle ABC \cong \angle A'B'C'$. Using SAS we can say that $\triangle ABC \cong \triangle A'B'C'$. In the case that $\angle ACB > 90°$ a similar argument applies.

$\angle ACB < 90°$ $\angle ACB > 90°$

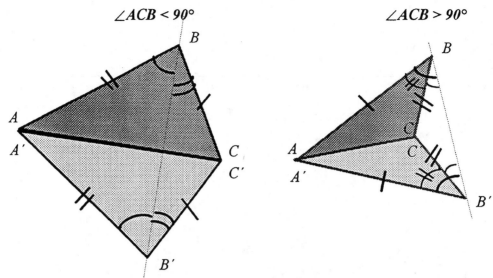

Given that ITT is true for all triangles on a sphere, and SAS is true for small triangles, this proof of SSS seems to hold on a sphere as well. Note, however, that among the definitions of small triangle for SAS there is one that is not sufficient for SSS: "a small triangle has all its sides less than 1/2 great circle." The counterexample in the figure below shows that no matter how small the sides of the triangle are, SSS does not hold:

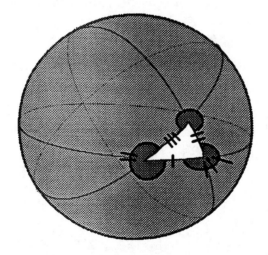

A large triangle with small sides

Thus, it is necessary to restrict the size of the angles for SSS to hold on a sphere. All of the other definitions that were discussed in Chapter V account for the case of the large triangle represented here.

2nd Proof:

Consider three straight segments such that it is possible to make a triangle using them. For one of those segments and for each of its endpoints, draw a circle that has an endpoint as its center and one of the sides of the triangle as its radius. We show this in the figure below. As we proved earlier in this chapter the circles must intersect twice. Thus, we have two possible triangles with corresponding sides congruent. But those triangles are necessarily congruent because the two circles each have reflection symmetry about the line containing AC because AC contains a diameter of each circle, and so the two triangles are mirror images of one another across this line.

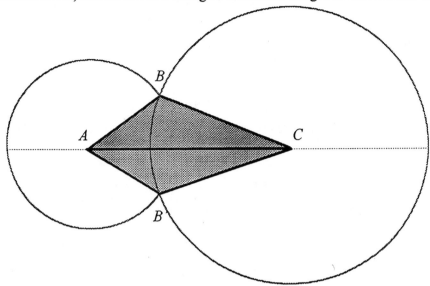

Example of Student's Work on Problem 13

Professor and T.A.   Student answer
comments:

Problem 13 - Ian Malcolm

Begin with one side and prove that, given two other sides, a triangle is defined uniquely:

Given AB: A ————— B   start constructing the triangle by

adding side BC: [diagram: A ——— B ✕ C]   Point B (a vertex) is defined.

Add side CA: [diagram: A — B ✕ C]   point C (also a vertex) is

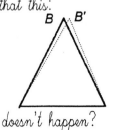

How can you be sure that this:

defined. This far, no angles are defined, but all three side lengths are fixed.

In order to form $\triangle ABC$, side AC must intersect AB at point A and then third vertex is defined.

Since the side lengths are fixed, the angles are fixed as well. This is

easier to see, perhaps, if we look at the following: [diagram: triangle with vertices A, B, C]

doesn't happen?

We move AC and BC together until they meet. There is a unique point in space, at which they intersect. The direction of AC relative to AB is fixed and the direction of BC relative to AB is fixed and then also $\angle ACB$ is fixed. So the triangle is uniquely defined.

Consider triangles: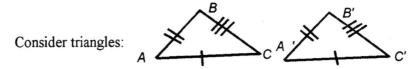

Since each part of each triangle is defined uniquely, in the same way, $\triangle ABC \cong \triangle A'B'C'$.

For all triangles?
Counter-examples?

This can also be proved by superimposing $\triangle ABC$ on $\triangle A'B'C'$. There is only one way to orient the sides, so, given that sides of $\triangle ABC$ and $\triangle A'B'C'$ are equal, we can superimpose $\triangle A'B'C'$ on $\triangle ABC$ (and vice-versa). So they coincide in the space and are congruent. This works on plane, sphere, cone and cylinder, given the right combination of rotation and reflection.

Response to comments:

Again, as long as the side lengths are fixed, the angles are fixed as well. In cases where the side lengths change, like your counter-example, we end up with a different triangle:

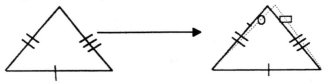

These two triangles are only congruent when ▭ $\cong$ || and O $\cong$ ‖ .

Why?

Does this convince you for the sphere also?

Otherwise, by moving the sides to get different triangles, we get different sides lengths.

I convinced myself of this by making triangles on paper and with sticks and things. I couldn't do it. By fixing the perimeter of one triangle, we have fixed the angles.

This is seen easily by making a triangle in a circle:

The arc swept out by α is fixed by sides AB and AC, and so $\angle \alpha$ is fixed. Likewise β and γ are fixed.

Good idea, but I'm not yet convinced. It seems to depend on 2 SSS triangles being in-scribed on the same circle.

By definition, every parameter of the triangle, we can see it's uniquely defined for all the cases on plane and sphere which I consider to be triangles.

Let's investigate counter-examples on the sphere... (for large triangle). Where two sides = 1/4 g.c. the third angle is ambiguous:

But for SSS we specify three sides. We get infinite number triangles. I guess this case is a triangle by my standards, so we have to exclude it.

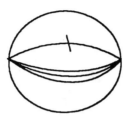

Where one side is =1/2 g.c., we also get an infinite number of triangles:

Look at:

Why so strong?

Proof?

For other large triangles, SSS holds, for my proofs, as long as we exclude the two sides less 1/4 g.c., SSS holds.

Response to comments:

I think my triangle inscription idea was lame. It was kind of an act of desperation anyway.

On the sphere, we get two triangles for a set of three sides.

These triangles have the same sides, so this is a counter-example. The main problem with this one is that the big triangle has finite area.

This one contradicts my restrictions that all angles of the triangle should be less than 180°. So, I don't consider it a

Good!

triangle.

There is no need to restrict the sides to less than 1/4 g.c.: you are right.

Look, I haven't the foggiest clue how to prove this, but I <u>know</u> that when you restrict all of your sides lengths, you can't do this:

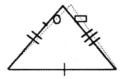

without changing side lengths. I tried to show this with my circle inscription, but I couldn't get it to work.

Let's look at it this way:

Assume we've connected all the points in some fashion:

Now we want to hold AC and move AB and BC around. By moving AB down (decreasing $\angle BAC$), we necessarily shorten BC because the point B must still lie on the area swept out by AB as it moves:

In other words, we can't do it. The reason is that we'd have to lengthen AB to change the area swept by AB in order to keep BC of constant length.

We can draw this: B must lie always on both of these two areas. As we can see if A and C are fixed, only B is the point of intersection for the two given lengths.

These areas also intersect again to give a congruent triangle below AC. This also works on the sphere.

PROBLEM 14. Angle-Side-Side (ASS)

Are two triangles congruent if an angle, an adjacent side, and the opposite side of one triangle are congruent to an angle, an adjacent side and the opposite side of the other?

Students will investigate ASS on the plane and then on a sphere. Let us look first at a counterexample of ASS on the plane:

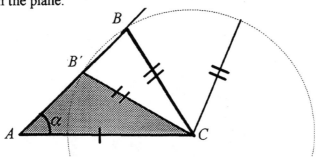

Students will probably come up with this counter-example. Here, the circle that has as its radius the second side of ASS intersects the ray that goes from A along the angle α in two places. One way to get around this might be to define α as a right angle, then:

ASS holds for right triangles (where the Angle in Angle-Side-Side is right).

Proof:
Consider a circle that has as its radius the second side of ASS, and consider the line on which the third side of the triangle lies. Given that the angle of ASS is a right angle, this line has to be perpendicular to the first side. The possible triangles that can have the third vertex on that line and on the circle (by the restriction of the length of the second side) are $\square \triangle ABC$ and $\triangle A'BC$. These two triangles are unique given that, in the plane, a line intersects a circle at most twice (see Chapter 7 and note that if the circle intersects the third side only once we do not have a triangle). Now we have to see if $\triangle ABC \cong \triangle A'BC$. Since C is the center of the circle, the segment BC extends to a diameter. This diameter is a line of reflection symmetry for both the circle and the perpendicular through A and A', so we can reflect $\triangle ABC$ about it. It follows that A must reflect onto A', and thus $\triangle ABC$ reflects onto $\triangle A'BC$, and so we have $\triangle ABC \cong \triangle A'BC$.

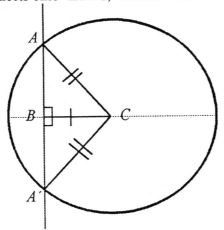

Many students are likely to prove that $\triangle ABC \cong \triangle A'BC$ in the following manner: Given that $\triangle ACA'$ is isosceles, ITT holds, and so given that CB is perpendicular to the base of $\triangle ACA'$, it must coincide with the bisector of $\triangle ACA'$. [Note that this last property is true on the plane but not on the sphere (in the case the circle is a great circle). Thus it is appropriate to ask the students why it is true on the plane.] Then using SAS we have $\triangle ABC \cong \triangle A'BC$.

We have just proven the Right-Leg-Hypotenuse Theorem (RLH) for the plane, which can be expressed in the following way:

If the leg and hypotenuse of one right triangle are congruent to the leg and hypotenuse of another, then the triangles are congruent.

At this point, students might conclude that since the theorems used to prove RLH are true for small triangles on a sphere, it suffices to choose any of the definitions of small triangle that worked for SAS to prove RLH on a sphere (given that ITT is true on a sphere for all triangles). But there *are* small triangle counter-examples to RLH on spheres! The following counterexample will help students to see some ways in which spheres are intrinsically very different from the plane.

In the figure below we can see that the second leg of the triangle intersects the geodesic that contains the third side an infinite number of times. So on a sphere there are triangles which satisfy the conditions of RLH although they are non-congruent.

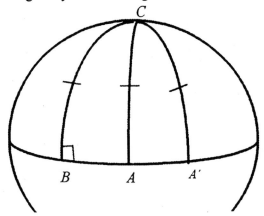

$$CB = CA = CA' = \text{1/4 of a Great Circle}$$

Let us look at possible ways to get out of this situation:

On a sphere, RLH is true when all the sides of the triangle are less than half a great circle and the hypotenuse is not equal to 1/4 of a great circle.

Proof:

Consider a right triangle in an open half-hemisphere (this will force each of the sides to be less than 1/2 great circle). Now consider the circle with radius equal to the second side (which we know by hypothesis must be smaller or larger than 1/4 great circle) that has as its center the vertex between the two given sides. The circle contains all the possible third vertices of the triangle, and it intersects the third (non-given) side of the triangle only once (note that all the sides have to be less than half a great circle). Thus, for any triangle with the same given angle and the same first given side, it must be that its hypotenuse and third vertex coincide with those of the first triangle.

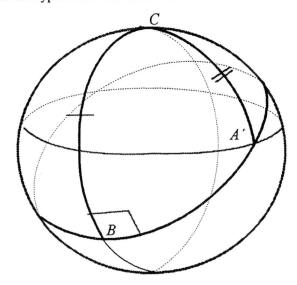

Another result:

ASS is true for triangles for which all angles are less than or equal to 90°.

Proof:

This requirement will result in ASS congruence because any construction of a non-congruent triangle with the given angle and first given side coinciding will result in an angle larger than 90°.

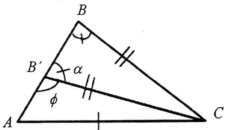

In the figure above, angle ϕ is greater than a right angle because, first, ϕ and α together constitute a straight line (or 180°) and, second, $\alpha \cong \angle ABC$ (using ITT), and by definition $\angle ABC$ is acute. So, it follows that ϕ must be larger than a right angle.

A further result:

ASS is true for those plane triangles and small, spherical triangles which have the second given side longer than the first given side (or AS_1S_2 is true if $S_1 < S_2$).

Proof (without words):

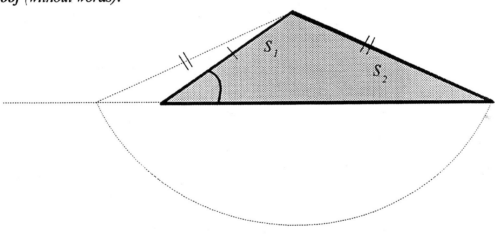

$$S_2 > S_1$$

Another Proof:

Consider $\triangle ABC$ as $\triangle AS_1 S_2$ with $S_1 < S_2$. Also consider the two circles of center B and radii S_1 and S_2, respectively. The circle S_1 is contained in circle S_2 because $S_2 > S_1$ and they share the same center. Let C, C' and C'' represent possible choices for S_2 such that the angle of ASS is an acute angle, a right angle or an obtuse angle, respectively. Now the rays AC, AC' and AC'' intersect the S_1 circle at most twice. But they intersect the S_2 circle only once since S_2 contains S_1 and all the rays emanate from the same point A which lies on circle S_1. It follows then, that for each angle, the triangle is unique.

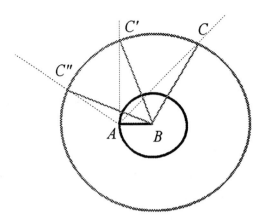

Another proof of $S_2 > S_1$

Example of Students' Work on Problem 14

Professor and T.A. comments:

Student answer:

<u>Problem 14 - Justin Collins</u>

ASS is true on the plane because, as in the other cases, ASS uniquely determines the orientation and length of the second and third sides. I like to think of it this way: Imagine that you have two sides of fixed length, attached at one end-point. Also imagine that the second side is attached to the first with some sort of pivot joint, so that it is free to rotate around the point of intersection, while the first side is fixed:

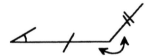

Draw the third side so that it joins the first side at a given angle and extends to infinity. Now rotate the second until it intersects the third side, but does not pass through it, i.e. no part of the second side "sticks out" on the other side of the third side. Allowing the second side to pass through would change the length of that side of the triangle:

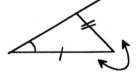

Behold:

$\triangle ABC \neq \triangle AB'C$

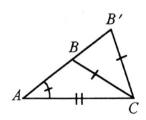

If the second side was not fixed at a particular length or could pass through the third side, then there would be an infinite number of intersections. Because the second side is of a fixed length, however, there is only one place that it can intersect the third side. This determines uniquely the length of the third side, and the orientation of the second side. *Note: If the given angle is too large, or if the lengths of the other two sides are not appropriate, i.e., the first side is too long or the second side is too short, then the given ASS will not form a triangle at all.*

With the cone, we once again run into the same problem with the cone point. Because you can move a triangle around on the cone without distorting except at the cone point, once again the cone point cannot be in the interior of the cone. But the problem is slightly different this time. If you allow the triangle to wrap around the cone point, then, as before, the third side is shortened:

Now, if we move this construction down, it is possible that we can not even make a triangle out of these sides:

Another difficulty can occur when the triangle wraps around the cone point. It is possible for the third side to have two different lengths, depending on the length of the second side:

Therefore we once again have to use the small triangle definition in order to keep the cone point out of the interior of the triangle. With the cone point out of the picture, we can see that the sides of the triangle are restricted to one unique intersection. There is no "optional" route that the third side can take. No matter what the angle, there is only one place that the two sides can intersect because of the restriction of the second side.

For the cylinder, problems can occur with the angles formed by the second side. If the second side is sufficiently long, we can manipulate the length of the third side and the angles that are involved with the second side:

We can remove these problems by confining the triangle to a one-sheeted covering. This removes the chance of there being an infinite amount of geodesics with the same length as the second side. If the second side can't wrap around, then there is only one unique path, i.e. multiple spirals of the same length are eliminated.

For the sphere, problems can occur if the second side is "sufficiently" long. For any given ASS, there are certain combinations of angles and sides which can produce two different triangles:

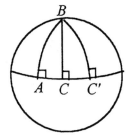

$\Delta\,ABC \neq \Delta\,AC'C$

not enough

Limiting the third side to less than 1/2 great circle will rule out this second intersection, which caused the ASS to be false. As in the plane, there will be only one unique intersection possible between the third side and the second side.

Response to comments:

I see now that dealing with ASS on any surface is quite a mess. I was drowning in a sea of triangles during my first write-up of the ASS case, so I didn't catch what would happen if you had the right combinations of sides and angles. I've played with a new and improved model and discovered that the only time there is a unique side is if the angle between the second and third sides is ninety degrees:

What if this angle is 90°?

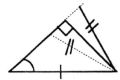

Response to comments:

If you look a little further, you can see that, on the plane, ASS works for right triangles. If you are a given triangle, you can see that given a sufficiently long enough second side (second side > first side), then there is only one intersection. On one side of the point, the second side is too short, on the other side is too long:

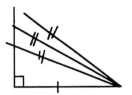

On a sphere, however, even right triangles don't work. (No, sometimes, see below). As the diagram that Eduarda drew on my papers shows, two sides and a right angle don't even determine a unique triangle, no matter what length:

What happens to the geodesics when they intersect at 90°? That is, what is the position of the geodesics when they intersect at 90°?

If you keep the legs less than 1/4 of a great circle, however, you get right triangles that satisfy ASS:

Good

These right triangles obey ASS for the same reasons they do on the plane. We can also allow legs between the lengths of 1/4 and 1/2. First consider the case where the first leg is less than 1/4 of a g.c. and the second (given) side is shorter than the first:

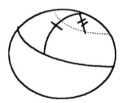

("latitude" contained in a geodesic)

If we rotate the second side around, we can see that it traces out a latitude line. Therefore, if the third side is at 90°, the latitude line will be completely contained in it and they will never intersect. Next consider the case where the first leg is less than 1/4 of a g.c., but the second side is greater than the first. If you rotate the second leg around, you will get a latitude line that will intersect the third side twice:

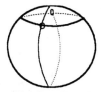

If you limit all of the sides to less than 1/2 great circle, however, this intersection will not be allowed.

Nice.

A similar argument works if the first side is longer than 1/4 g.c., but less than 1/2. You get latitude lines, traced out by the second sides, that have the potential to intersect the third side twice. As long as you limit the sides to less than half a g.c., only one intersection is possible. Therefore, right triangles obey ASS on a sphere and on a plane if the legs of the triangle are less than 1/2 g.c., and legs of length 1/4 g.c. are not allowed.

For a cylinder the same problem we had with the plane occurs. Right triangles will satisfy this requirement, as long as we rule out the optional longer sides that helix around the tube of the cylinder:

As long as we eliminate these longer sides by not allowing the triangle to extend past an open one-sheeted covering, ASS is true for right triangles on the cylinder. The same holds for a cone. As long as the cone point is not included in the interior, right triangles obey ASS.

Chapter 8

Parallel Transport

Problems 15, 16 and 17 allow students to further develop the notion of *parallel transport* that was introduced in the previous chapter. Students may choose to use Problem 15 to solve Problem 16, or conversely, they may opt to use Problem 16 to solve Problem 15. These problems will enable them to solve Problems 18 and 19.

PROBLEM 15. *Euclid's Exterior Angle Theorem (EEAT)*

Any exterior angle of a triangle is greater than each of the opposite interior angles.

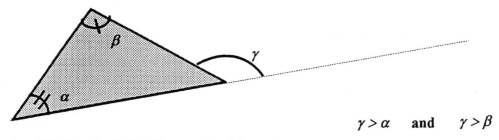

$$\gamma > \alpha \quad \text{and} \quad \gamma > \beta$$

Look at EEAT on Both the Plane and a Sphere

The following hint (which is found in Euclid's writings) may be useful, although it is not necessary to use this hint to solve the problem: Draw a line from the vertex of α to the midpoint, M, of the opposite side, BC. Extend that line beyond M to a point A' in such a way that $AM \cong MA'$. Join A' to C.

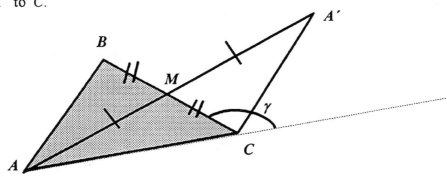

Euclid's Hint

Let us look at a proof of EEAT for triangles on the plane using Euclid's hint:

Consider $\triangle ABM$ and $\triangle A'CM$. By construction we have that $BM \cong CM$ and $AM \cong A'M$. By VAT (Chapter 3) we have $\angle BMA \cong \angle CMA'$, and thus by SAS (Chapter 5) we have $\triangle ABM$

$\cong \Delta A'CM$. Consequently, we have that $\angle MCA' \cong \beta$. But $\angle MCA'$ **is inside of** γ, and so $\gamma > \beta$.

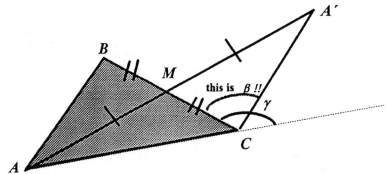

$$\beta \text{ is Inside } \gamma$$

In a similar manner, using Euclid's hint on side AC, we can prove that $\gamma > \alpha$. We show this in the following figure:

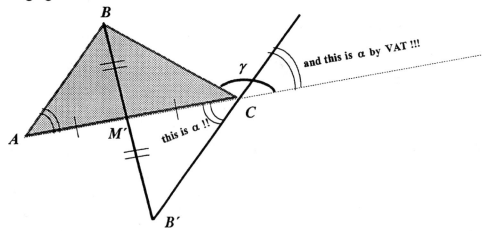

$$\alpha \text{ is Inside } \gamma$$

Let us look at another proof of EEAT on the plane that does not involve VAT or properties of SAS.

Consider the image of a triangle through half-turn symmetry around the midpoint of the side opposite to angle α. Since the new triangle is obtained by a symmetry property of a straight line, we have that the original triangle and its image are congruent. Thus, $\angle ABC \cong \angle A'CB$, and since $\angle A'CB$ is inside of γ, we have that $\angle ABC$ has to be inside of γ. A similar half-turn symmetry using the midpoint of AC will produce the desired result for α.

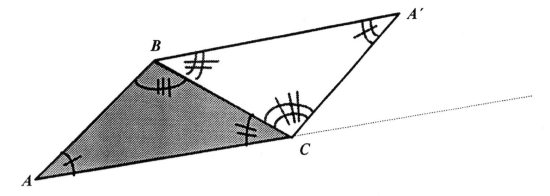

A Proof of EEAT on the Plane by Symmetry

Both of these proofs seem to work on spheres. The first proof seems to work for the same class of triangles for which SAS holds on a sphere. The second proof seems to work for all spherical triangles given that we can always rotate a triangle in the manner described above. Yet, EEAT is not always true on a sphere, even for small triangles. Let's look at a counter-example:

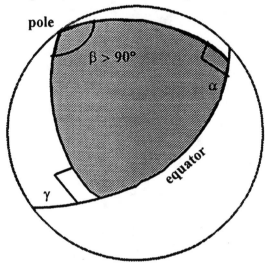

EEAT is not true on a sphere

What is wrong with the proofs we have given for EEAT on a sphere? You might suggest to the students that they try out Euclid's hint on a small spherical triangle and see what happens. In the figure above Euclid's hint produces a median that, when extended, will intersect beyond the side which we extended to form γ. In this case, $\beta > \gamma$.

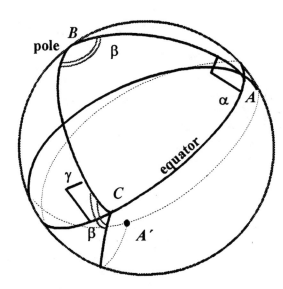

Euclid's Hint on a Sphere: For this triangle, γ is inside of β !

On the plane, we would not expect the line containing the median of the side opposite to angle α to intersect the line containing *AC* at any other place than *A*. Since lines on a sphere intersect in two places, we can avoid the situation illustrated in the figure above by requiring that the triangles have a median that is less than 1/4 of a great circle. If every median is less than 1/4 of a great circle, then every extension will be less than half of a great circle, and there will not be a second intersection with the line containing *AC*. Consequently, β will be always inside of γ. And conversely, note that if the median is not less than 1/4 great circle, then EEAT definitely fails. Thus we have shown that in the case:

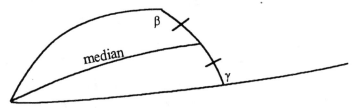

The interior angle, β, is less than the exterior angle, γ, if and only if the median has length less than 1/4 great circle.

More proofs of EEAT will be illustrated after the discussion of Problem 16. Some of those proofs will use Problem 16, while others will be based on insights gained from Problem 16 involving lines which cut at congruent angles.

Example of Students' Work on Problem 15

Professor and T.A. comments: *Student answer:*

Problem 15 - Eileen Hannigan

Given: 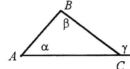   prove γ>α and γ>β .

Proof: Draw a line *AD* such that *AD* intersects *BC* and the two segments bisect each other:

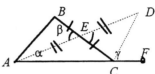

Then $AE \cong ED$, $BE \cong EC$ by definition of bisector and $\angle BEA \cong \angle DEC$ by the VAT. Thus by SAS, $\triangle AEB \cong \triangle DEC$.

By corresponding parts $\angle ECD \cong \beta$. However, $\gamma \cong \angle ECD + \angle DCF \cong \beta + \angle DCF$. Since $\angle DCF$ must be > 0: $\gamma > \beta$.

Since $\beta \cong \angle ECD$, *AB* and *CD* are two lines cut by a transversal such that alternate interior's are congruent. Thus $AB \parallel CD$. *AB* and *CD* are also cut by the transversal *AF*. Thus the corresponding angles α and $\angle DCF$ are congruent.

However, $\angle\gamma \cong \angle DCF \cong \angle BCD \cong \alpha + \angle BCD$ and then $\gamma > \alpha$.

<u>Sphere</u>: I drew Euclid's model on a sphere but I fail to see why, for small triangles, the proof would by any different than on the plane.

Check counter-example

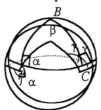

Response to comments:
EEAT on sphere:

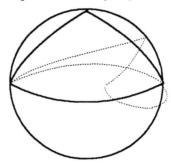

Does not hold for triangles where more than one side is equal to 1/4 g.c. In this case γ = α.

Does not hold for triangles where a side is > 1/2 g.c.

$\alpha > 180°$ and $\gamma < 180°$, so γ is not bigger than α.

Restrictions?

Response to comments:

So, EEAT holds on the sphere when all the sides of the triangle are less than 1/4 great circle. In this case we can see that my proof on the plane (using Euclid's Hint) works.

PROBLEM 16. *Symmetries of Parallel Transported Lines*

*Consider two lines, **r** and **r′** that are parallel transports of each other along a third line, **l**. Consider now the geometric figure that is formed by the three lines, one of them being a transversal to the other two, and look for the symmetries of that geometric figure.*

*What can you say about the lines **r** and **r′**? Do they intersect? If so, where?*

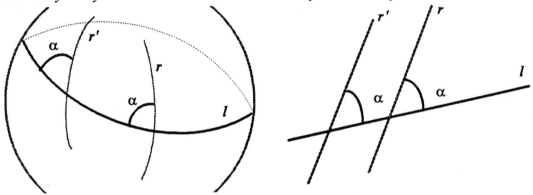

r′ is a parallel transport of r along l

The notion of parallel transport was already introduced informally in Chapter 6. In Problem 16, students have an opportunity to explore the concept further and prove its implications on the plane and sphere. They will also study the relationship between parallel transport and parallelism. Historically, Euclid and many other mathematicians avoided the use of a parallel postulate as much as possible for reasons which we will explore in Problems 20, 21, and 22. The students in our course have shown almost no inclination to use any parallel postulate but they have a strong inclination to use two of its consequences on the plane: *Any transversal of a pair of parallel lines cuts these lines at congruent angles* (Problem 20) and *The angles of any triangle add up to a straight angle* (Problem 23). The use of these results should be avoided for now, as they are both false on a sphere. We have been investigating the similarities of the plane and sphere and trying to use common proofs whenever possible.

Symmetries and Parallel Transport

What is meant by *symmetry* with regard to geometric figures? A transformation is a *symmetry* of a geometric figure if it transforms that figure into itself. That is, our focus is on the global symmetries of the figure. Here, we are looking for the symmetries on both the plane and a sphere:

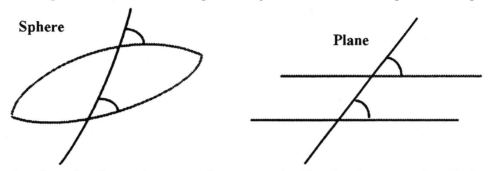

It is clear from the picture that on a sphere we are looking for the symmetries of a *lune* cut at congruent angles by a geodesic. A ***lune*** is a spherical region bounded by two half great circles.

Let us look first at the plane. Consider the midpoint of the transversing segment between the two lines. If we half-turn the whole figure about this point, we will get another figure that coincides with the original. Why is this true? Using the Vertical Angle Theorem and the fact that α and β add up to a straight line, we have the angle congruency expressed in the figure below. On the other hand, the rotation is done through the midpoint of *AB*, so a segment of *r* near *l* will be rotated onto a segment of *r*. If we now consider these segments as extended to the complete lines *r* and *r′*, then the rotated line *r* must coincide with the line *r′*. The property that two straight lines (or geodesics) must coincide if they have a segment in common follows from the symmetry properties we developed in Problems 1 and 2. The line *l* will coincide with itself since we are rotating it around one of its points and straight lines have half-turn symmetry.

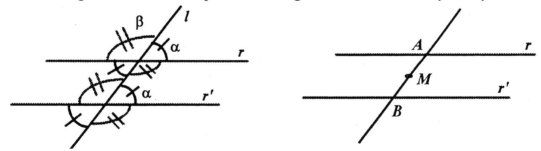

Half-turn symmetry in parallel transport

Note that since the geometric figure above has half-turn symmetry around *M*, what happens on the right-hand side of *M* <u>also</u> happens on the left side of it. Consequently, if the lines *r* and *r′* intersect on the right-hand side of the figure then by symmetry they will intersect again on the left-hand side. But, since lines on the plane do not intersect more than once, one may conclude that *r* and *r′* never intersect, that is, they are parallel.

One can also prove that *r* and *r′* do not intersect using EEAT. If the lines were to intersect at a point *C* then $\triangle ABC$ would have an exterior angle congruent to the interior angle α, which cannot happen in the plane. Thus, the lines must not intersect.

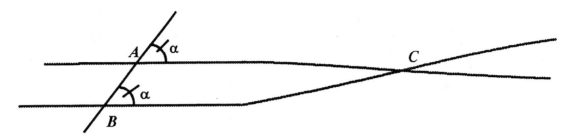

On the plane parallel transported lines do not intersect

Students seem to want to be able to parallel transport the line l simultaneously along the two lines r and r'.

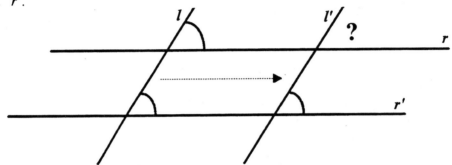

Parallel transporting the transversal

However, this is clearly impossible on a sphere, and so the possibility of doing so on the plane must be examined carefully. We can certainly parallel transport l along r' to the line l' (on any surface) and keep the three angles marked in the figure congruent. What is then needed is an argument that the angle between l' and r is the same as the other angles. This is true if we have already proved that the lines are parallel and if we know the result from Problem 16. In this manner, students might want to argue for some kind of translation symmetry. That is, *on the plane the line l can be parallel transported along the two lines simultaneously without changing its angle of intersection with either of the lines.*

Let us look now at the situation on a sphere:

Let P and P' be the intersection points of any two lines r and r' on a sphere (refer to the following diagram). Given that the transversal l cuts r and r' at congruent angles, we have the same congruency situation that we found on the plane. Consider $\triangle APB$ and $\triangle AP'B$. These two triangles are congruent by ASA. In particular, the midpoint of AB is an equal distance from P and P' (since the triangles are congruent they have the same median), and so it coincides with the center of the lune formed by r and r'.

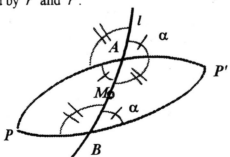

Using the same argument for the plane and sphere, we conclude that each of the respective geometric figures has half-turn symmetry. In the case of a sphere, we found a relationship between the intersecting points of a lune and a transversal of the lune that cuts at congruent angles:

If a transversal of a lune cuts at congruent angles then it will go through the center of symmetry of the lune.

Let us look at another proof of the above result. If l cuts r and r' at congruent angles then consider the midpoint, M, of the segment of l between r and r'. Drop a perpendicular from M to r. Then one can form $\triangle AMA'$ and $\triangle BMB'$. Using VAT and ASA these two triangles can be shown to be congruent. Thus, $B'M \cong A'M$, and so M has to be the center of the lune:

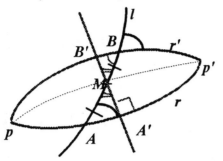

The converse of the above statement, which is investigated in Problem 17, is implicit in this proof.

Example of Students' Work on Problem 16

Professor and T.A. comments:　　　*Student answer:*

Problem 16 - Jon Howard

<u>Plane</u>: It has 180° rotational symmetry about a point L halfway between L_1 and L_2.

L_1 and L_2 are parallel.

You can use this in your argument below.

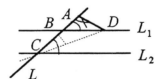

If L_1 and L_2 are not parallel, they meet at a point. Connect this point with any point on L above L_1. Then by SAA the two triangles are congruent. Then by CPCTC, $AB \cong AC$. But AB is not congruent to AC (assuming L_1 and L_2 are not the same line). So L_1 and L_2 do not intersect. So L_1 and L_2 are parallel.

OK, if you finish #18 without using

Above, on the plane, you concluded that they were parallel, based on an argument involving angles...

<u>Sphere</u>: The given situation happens only in very restricted situations. If L_1 and L_2 are considered longitudes, then the equator must intersect L precisely halfway between L_1 and L_2, and L cannot go through the poles.

L_1 and L_2 are not parallel. Since they are straight lines, then they must intersect at two points (assuming non-collinearity). Hence they are not parallel.

It does, however, have 180° symmetry.

Response to comments:

Sphere: Diagram to help see the first paragraph of the original:

And then?

What point is

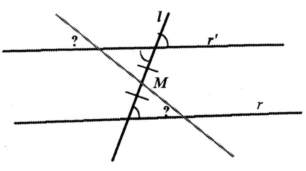

If the transversal is the equator, obviously the two angles will be congruent. If L is tilted, the angle changes the same amount in both directions, due to symmetry. Anywhere else, the symmetry does not exist, and so the situation described in the problem does not exist.

From the second paragraph: The usual definition of parallel is that straight lines do not cross. Since these lines are defined as straight, they must cross on a sphere, and hence are not parallel. You don't need any argument about angles for this. You could say instead that any two straight lines are parallel, but regardless, it has the same effect. Two straight lines must cross on a sphere at the poles.

PROBLEM 17. *Transversals Through a Midpoint*

*If two geodesics **r** and **r′** are parallel transports along another geodesic **l**, then they are also parallel transports along any geodesic passing through the midpoint of the segment of **l** between **r** and **r′**.*

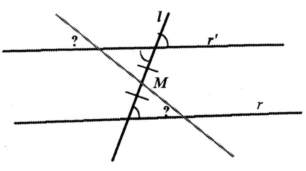

*Does this work for both the plane and sphere? On a sphere are these the only lines which will cut **r** and **r′** at congruent angles? Why?*

Let the intersections of the new line with the lines r and $r′$ be $A′$ and $B′$. We have to prove that $\angle AA′M \cong \angle BB′M$. Consider the triangles $\triangle A′MA$ and $\triangle B′MB$. Using VAT and ASA (note that both triangles are contained in the lune and consequently all of their sides are less than half great circle), we can prove that $\triangle A′MA \cong \triangle B′MB$. But then $\angle AA′M \cong \angle BB′M$.

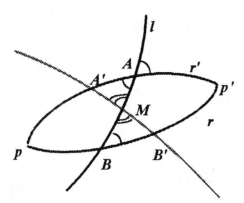

All lines that cut at congruent angles go through the center of the lune

Example of Students' Work on Problem 17

Professor and T.A. comments: *Student answer:*

Problem 17 - Shawn Haarer

Show that a parallel transversal that passes through the midpoint p will cut $L1$ and $L2$ at congruent angles. (The <u>converse</u> of 16.) (We used and proved this for a sphere on Problem 16, but let me recap for both the plane and a sphere.)

<u>Sphere:</u>

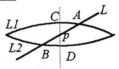

If [AB passes through] the midpoint of the equator, then we have congruent triangles ACp and BDp by ASA. Hence the angle of L at B is congruent to the angle of L at A by corresponding parts. We can also show this by the same reflection argument used above. Since pB is a reflection of pA through p and pC is a reflection of pD, then we can use half turn symmetry to lay pA on top of pB and pC on top of pD. Then $\angle CAp$ will lie on top of $\angle DBp$.

<u>Plane:</u>

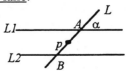

The symmetry argument used above will certainly work on a plane. Further, since the lines $L1$ and $L2$ are forever equidistant (unlike the sphere), there is no equator on which the point must lie.

Proof of EEAT using Problem 16

Let us now look at a proof of EEAT using Problem 16:

Parallel transport angle α so that its vertex coincides with vertex C of triangle ΔABC. On the plane the line AB and its parallel transport never intersect, and so α will be inside of γ. On a sphere they will intersect, and so their place of intersection must be determined. If the median from B is less than 1/4 of a great circle then the intersection will be inside the lune AB-BC, and thus α will be inside of γ.

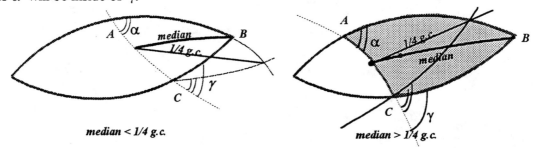

median < 1/4 g.c.   *median > 1/4 g.c.*

The notion of parallel transport gives you a way to check parallelism. Even though parallel transported lines intersect on a sphere, there is a feeling of *local parallelness* about them. The issue is not whether the lines ever intersect but if they can be cut at congruent angles at a certain point. That is, whether the lines are *locally parallel*. You may choose to avoid definitions of parallel that do not give you a direct method of verification. Definitions such as:

- *Parallel lines are lines that never intersect,* or

- *Parallel lines are lines such that any transversal cuts at congruent angles,* or

- *Parallel lines are lines that are everywhere equidistant*

are definitions that are not human in the sense that they can not be verified directly. In Problems 20 through 25, students will compare the notions of parallelism and parallel transport, and in the process they will learn something of the history and philosophy surrounding parallel lines.

On the plane, straight lines are parallel without ever intersecting; on a sphere, straight lines are locally parallel and converge symmetrically; and on a hyperbolic plane, straight lines can be locally parallel and diverge symmetrically (and thus not intersect). However, there are also pairs of "asymptotic" lines on the hyperbolic plane which do not intersect (and are thus parallel by the usual definition) but which are not parallel transports of each other along any transversal. If you look at a cone with cone angle larger than 360°, you also can find two non-intersecting lines which are not parallel transports of each other. These examples, together with Problems 15 through 25, should give you the understanding that parallelism is not just related to non-intersecting lines.

Chapter 9

SAA and AAA

Problems 18 and 19 ask students to investigate two triangle congruence properties. The first is "Side-Angle-Angle," referred to as SAA, and the second is "Angle-Angle-Angle," or AAA.

PROBLEM 18. *Side-Angle-Angle (SAA)*

Are two triangles congruent if one side, an adjacent angle, and the opposite angle of one triangle are congruent, respectively, to one side, an adjacent angle, and the opposite angle of the other triangle?

You may want to suggest a general strategy for making two triangles coincide as much as possible. Let us look at this for SAA:

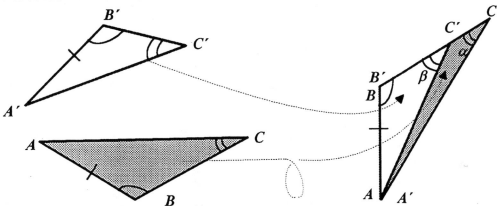

Most students will not feel comfortable with the above diagram. Most will want to argue that it cannot happen. Some may want to argue that α and β cannot be congruent angles, that it is not possible to construct such a figure. Most students will come to understand that this diagram is, indeed, impossible on the plane, but it is important they also see that this situation is possible on a sphere, as is shown in the next figure:

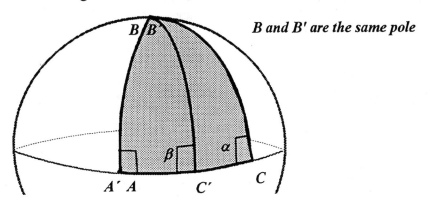

B and B' are the same pole

Some students may feel certain that it is the only counterexample of SAA on a sphere. You can show students that there are in fact other counterexamples for SAA on a sphere by using the notion of parallel transport, which is the underlying idea for such counter-examples.

With the help of parallel transport, one can construct another counterexample for SAA on a sphere. For example, as we show in the following figure, take a short straight segment that makes an angle α with a geodesic g, and parallel transport that straight segment along the geodesic g. Consider the segment and its image after transport, and extend those two segments to two geodesics g_1 and g_2. Since every pair of geodesics on a sphere intersect, we can choose one of the intersections of g_1 and g_2, namely B, and draw a line from that intersection point to A. The two resulting triangles provide one counter-example for SAA on a sphere, but the technique we have used also leads to infinitely many others.

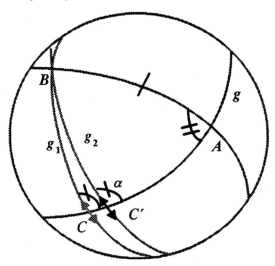

Many Counterexamples for SAA on a Sphere

$\triangle ABC$ and $\triangle ABC'$ satisfy the conditions of SAA, and yet they are not congruent. Is there a restriction we can make on triangles on a sphere so that SAA *is* true? Students will be able to answer this question using Problems 15 and 16 from the last chapter.

Some students might want to prove SAA on the plane using the result: *The sum of the interior angles of a triangle is 180°.* This result will be proven later for the plane, but it does not hold on spheres. Thus, students should avoid using this result and, instead, use the more useful concept of parallel transport. This suggestion stems from our desire to see, as much as possible, what is common between the plane and sphere. In addition, before we can prove that the sum of the angles is 180°, we will have to make some additional assumptions on the plane which are not needed here.

Proof of SAA

Let us now see how Problems 15 and 16 can shed some light on Problems 18 and 19. Problem 18 on the plane can be solved using Problem 16. Consider the situation illustrated in the figure below:

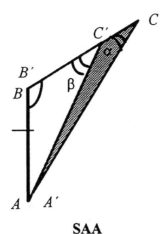

SAA

If this situation were possible (that is, suppose AC and AC' are not congruent) then the lines l and l' that contain the segments AC' and AC, respectively, would be cut at congruent angles by the line that contains the segment BC. Consequently, on the plane, they would not intersect by what was proven in Problem 16. Since they do intersect at A, the situation in the figure above cannot happen. So, $AC' \cong AC$, and we have proved:

SAA is true for all triangles on the plane.

Looking at the same situation on a sphere, we conclude that:

SAA is true on a sphere for all triangles with sides less than 1/4 great circle

Proof:

Consider $\triangle ACC'$ in the above figure. Since all of the sides of $\triangle ACC'$ are assumed to be less than 1/4 great circle, we can use EEAT and conclude that $\angle BCA$ is less than $\angle BC'A$. But this is not possible considering that they are given as congruent. Thus, when all sides are less than 1/4 great circle, then the sides AC and AC' should coincide and the triangles are congruent.

Another proof:

The results from Problems 16 and 17 show us that when a transversal cuts two geodesics at congruent angles, it must go through the center of the lune formed by them. Also, the transversal forms two congruent triangles from the lune, each of which has at least one side greater than or equal to 1/4 great circle. Consequently, with the assumption that all sides of the triangle are less than 1/4 great circle, the transversal BC cannot cut the sides AC and AC' at congruent angles:

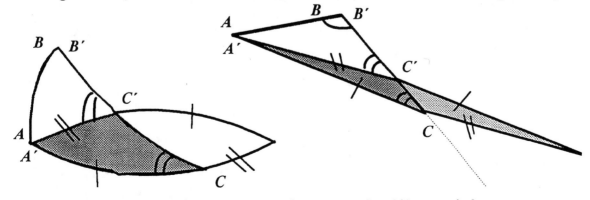

When the sides are greater than or equal to 1/4 great circle

Example of Students' Work on Problem 18

Professor and T.A. comments: Student answer:

Problem 18 - Kevin Newman

Why? What is your definition of

AB is the given side, angle ABC is given and angle ACB is given. Let's assume (for a contradiction) that there is more than one way to connect A and C — AC'. Since ACB is fixed, angle AC'B is the same as angle ACB. However, if angle AC'B and angle ACB are the same, then AC is parallel to AC'. This is a contradiction, however, because AC and AC' meet at A and are, therefore not parallel. So there is no C' different from C; AC is uniquely determined by AB, angle ABC and angle ACB.

Find a counter-example on the sphere and see what can be taken from your proof to the sphere.

This proof won't work on spheres, however, because there are no parallel lines on a sphere — any two straight lines will always cross.

Response to comments:

I looked at SAA on the plane and you asked me what my definition of parallel was. On the plane, my definition of parallel is that two lines no matter how far they are extended will never meet. Also, and as a result, they are always the same distance apart at all times — i.e. the

Proof?

distance you would have to translate one segment of the line to make it collinear with the other line is the same for all line segments (you can choose any one you like). If the lines meet at some point, no translation alone can ever make the two collinear (the lines are not separated by a constant distance — it goes from zero to infinity!).

I wonder if I should come up with a definition of parallel for the sphere because it seems as though the same properties might be useful even though the lines cross (actually, isn't spherical geometry the result of violating the postulate that parallel lines never meet?) each other — maybe all lines are parallel — two great circles can always be translated to lie on top of one another. Maybe that's enough of a definition and we can say that on the plane, the lines also have the property that they never cross each other.

What about the

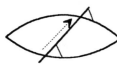

See the problems after problem 20.

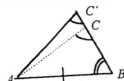

So let's look at the sphere. SAA does not hold on the sphere because parallel lines can meet — so AC' and AC can meet at A and yet be parallel and distinct. Therefore, it is possible to have angle ACB and angle $AC'B$ the same and BC and BC' different. So this is no good.

Here is another counter example:

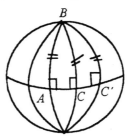

Can you make restrictions? Is **SAA** *true for* <u>*very*</u> *small triangles?*

Let BA be the given side and the two given angles be 90° — we already know that this forms many triangles from looking at it under ASS.

SAA cannot work in general. Given a side and two angles, the angles form two great circles which cross at two places, and the angles at their crossings are the same by the vertical angle theorem, so it is totally unclear how to pick the third side.

Here is a thought on the parallel lines transversing thing: I think that if you take two lines and cut them with a transversal, you can get the congruent angle property we see on the plane — if the transversal is of a special type. Take your two parallel lines and go 1/4 g.c. along each from the point where they cross. Those two points are their midpoints and a line joining their midpoints would be their equator if you take the first two points as poles. Clearly, the equator is perpendicular to both and the transversal thing holds. Now, go away from the midpoint in each line — in opposite directions (so, north on one and south on the other if you want to think of the perpendicular line as the equator) for equal distances. Connect those two points with a transversal. I believe that this new transversal has the properties of general transversals in the plane. The points at which the transversal cuts the two lines must be symmetric in terms of the distance from the equator in opposite directions.

Good! Use for prob-lems 16 and 17.

[This student had not yet done Problems 16 & 17]

Well, if in the first counterexample, the length of AC is limited to less than 1/4 great circle, then the symmetry property cannot exist because in order for a transversal to generate the angle congruence property and one of the intersections is less than 1/4 great circle, the other must be greater than 1/4 great circle by the same amount.

If AB is less than 1/4 great circle and angle ABC is less than 90° I think we guarantee that AC is less than 1/4 great circle. Take $AB = $ 1/4 g.c. and angle $ABC = 90°$ then we have the case of the second counter-example, which is bad. If AB is less that 1/4 g.c. and angle ABC is less than 90° then clearly the length of AC decreases to less than 1/4 g.c. If angle ABC becomes greater than 90°, AC is greater than 1/4 g.c. but so is AC' so the transversal property does not hold.

Other situations which will produce the same property: $AB < 1/4$ g.c. and angle $ABC < 90°$, $AB > 1/4$ g.c. and angle $ABC > 90°$. And while we're at it, let's keep everything on one hemisphere not counting the boundary.

Under those conditions, I think it could work. Simpler to say that all lines must be less than 1/4 g.c. long.

For SAA on cones and cylinders, it's the same as the plane when you use normal definition of small.

PROBLEM 19. Angle-Angle-Angle (AAA)

Are two triangles congruent if their corresponding angles are congruent?

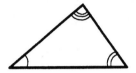

When trying to construct two triangles that satisfy the conditions of AAA, students will realize that they can construct two non-congruent triangles with corresponding angles congruent.

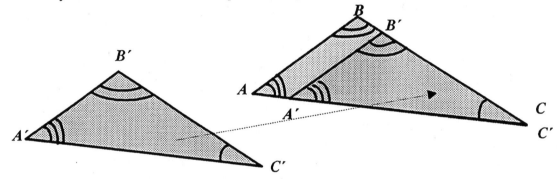

A counter-example to AAA on the plane

Parallel transport shows up in AAA in basically the same way it did in SAA, but here it happens simultaneously in two places. Students will recognize that in this case parallel transport produces similar triangles that are not necessarily congruent. However, since the objective is to look for a common argument for the plane and sphere, and since there are no similar triangles on a sphere, students should be encouraged to prove why such a construction is possible and why the triangles are not congruent. The construction seems intuitively possible, but students should justify *why* is it a counter-example. They will need properties of parallel lines such as in Problem 20. Students can be encouraged to assume these properties now as long as they make sure they do not use AAA when proving them later. Problem 37 will deal with the properties of similar triangles on the plane.

Ask yourself and your students: Is it possible to make the two parallel transport constructions shown in the above figure and thus get two non-congruent triangles? Suggest to students that they try to make such constructions and that they study the following situation on a sphere:

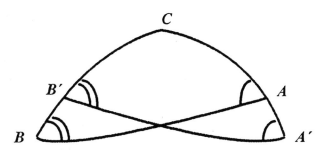

Does this produce non-congruent triangles on a sphere?

Proof of AAA

When we try to make two triangles that have corresponding angles congruent coincide as much as possible by using isometries, we are faced with one of the following two situations:

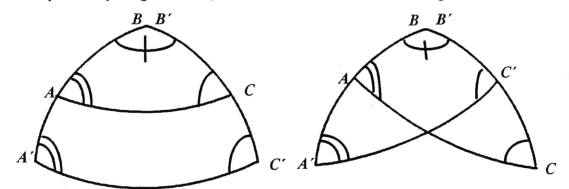

Case 1: both sides are shorter in one triangle than in the other.	**Case 2: one side of one triangle is shorter and the other side is longer**

Case 1 is possible on the plane, and it produces similar triangles. However, at this point students will need to use Problem 20 to be able to prove the existence of the construction. In a later chapter we will return to a discussion of similarity.

In Problem 16 we saw that if two lines are cut at congruent angles on the plane, then they do not intersect. Thus, Case 2 is not possible on the plane because the lines AC and $A'C'$ are cut at congruent angles by, for example, BC, but do intersect.

Thus,

AAA is false on the plane.

Let us look at these situations on a sphere. You may find that half of the class claims that AAA is true for every triangle on a sphere while the other half claims that AAA is not always true on a sphere. You might spend some time in class leading a debate about AAA on a sphere.

At this point, after having studied all of the congruency properties, each student will have a notion of triangle on a sphere and a notion of congruency. If a student includes intersecting triangles in their class of spherical triangles, then counter-examples can be found for Case 1:

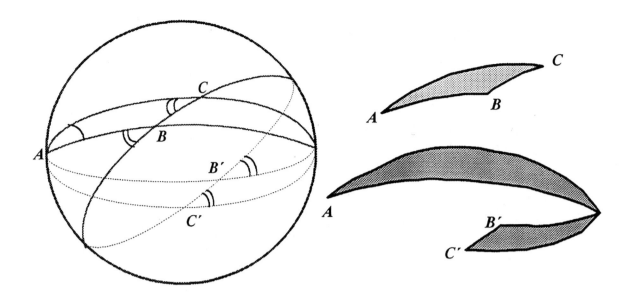

AAA does not hold when self-intersecting triangles are allowed

If a student allows triangles with straight angles, then Case 2 becomes an interesting counter-example that challenges the notion of congruency: The midpoints of the straight segments AA' and CC' are midpoints of the lunes lying on opposite hemispheres formed by the geodesics AC and $A'C'$. Thus, they must be opposite poles. This makes $\angle ABC'$ a straight angle because, given the angle congruency, A', A, C', C and B lie on the same great circle. So, the position of vertex B is arbitrary, and one can obtain two triangles that are congruent as geometric figures (given that they can be made to coincide) but which may not be congruent as triangles (because the vertices B and B' may not coincide).

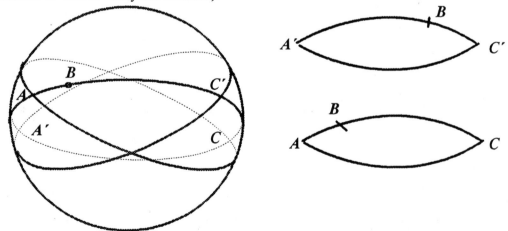

The geometric figures are congruent. What about the triangles?

Ask yourself: Is this notion of congruency the same as that which we have been using up to now? Or were we using something more specific? In fact, we were using congruency between two triangles to conclude that the *corresponding* sides were congruent. We were not only making the triangles coincide as geometric figures but we were also requiring that the corresponding vertices coincide. Some students will argue that AAA holds for all the triangles on a sphere because they do not consider self-intersecting triangles or triangles with straight angles to be valid triangles. Note that at this point the students seem to have *a common notion of triangle for the plane and sphere.*

In any case, we can conclude that:

AAA is true on a sphere for small triangles and for all non-intersecting triangles with no 180° angles.

Example of Students' Work on Problem 19

Professor and T.A. comments: *Student answer:*

Problem 19 - David Walend
AAA

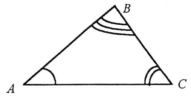

<u>Planes, cones</u> and <u>cylinders</u>:
AAA never works: The scale isn't locked. I can make the triangle any size I want and the angles still have all angles equal.
<u>AAA</u> on <u>sphere</u>:

This works!

Do problems 15 and 16 first.

The measure of angles on sphere are linked directly to the radius of the sphere. This locks in the scale of the triangle. AAA works for all triangles on a sphere - no matter how big!

I really need to know exactly what the relationship between length of lines, radius, and angles is. Any ideas?

Response to comments:

What point is c?

I think I have a proof on the sphere for AAA after doing #15 & 16. Choose two polar points. Draw two 1/2 great circles. Cut them so that congruent angles are formed and use vertical angles theorem. This only works for D + E = 1/2 g.c..

OK!

I don't think this is going anywhere yet. I need to ponder this one out.

Response to comments:
AAA defines the intersection of 3 biangles:

How is this possible to construct on a sphere?

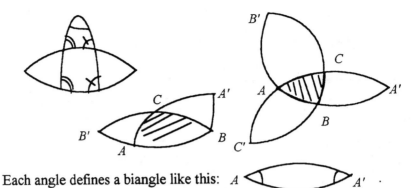

Each angle defines a biangle like this: A ◁▷ A' .

To join the first biangle to the second, let B lie on AA' and A lie on $B'B$. This way B', A, B and A' are all collinear.

Explain better your definition.

Now let C be the point where the other side of AA' intersects the other side of $B'B$. Slide biangle $B'B$ along AA' until angle C is what you asked for. For AAA to work you must restrict it to the inside of these three biangles. This probably over restricts AAA, but saves lots of trouble.

Response to comments:
AAA on sphere: Doesn't work on a sphere.

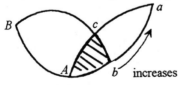

For these angles to be equal this line has to go through the center point.
If that is the case then it must look like :
and the larger item isn't even a triangle.

Sliding Biangles definition of AAA: Construct 2 biangles with corners of two of the specific angles. Slide them such that one side of each biangle is on a common great circle until they form the third angle. Does this limit the triangle to only one case?. Yes!

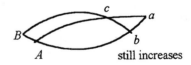

The other odd example:

isn't even a triangle at all.

Only works if this is a 2π angle - so

Chapter 10

Parallel Postulates

Parallel Lines on the Plane are Special

Up to this point we have not had to assume anything about parallel lines. No version of a parallel postulate has been necessary on either the plane or a sphere. We defined the concrete notion of parallel transport and proved in Problem 16 that on the plane parallel transported lines do not intersect. Now in this chapter we will look at two important properties on the plane:

If two lines on the plane are parallel transports of each other along some transversal, then they are parallel transports along any transversal. (Problem 20)

On the plane the sum of the interior angles of a triangle is always 180°. (Problem 23)

Neither of these is true on the sphere and both need an additional assumption for their proofs. The various assumptions that permit proofs of these two statements are collectively termed *Parallel Postulates*. Only the two statement above are needed from this chapter for the rest of the book. Therefore, it is possible to omit this chapter and assume one of the above two statements and then prove the other. However, parallel postulates have an historical importance and have a central position in many geometry textbooks and in many expositions about non-Euclidean geometries. The problems in this chapter are an attempt to help the students unravel and enhance their understanding of parallel postulates. Comparing the situations on the plane and to situations on a sphere is a powerful tool for unearthing our hidden assumptions and misconceptions about the notion of parallel.

Since we have so many (often unconscious) connotations and assumptions attached to the word "parallel" we find it best to avoid using the term "parallel" as much as possible in this discussion. Instead we will use terms like "parallel transport", "non-intersecting" and "equidistant".

PROBLEM 20. Parallel Transport on the Plane

Show that if l_1 and l_2 are lines on the plane such that they are parallel transports along a transversal l, then they are parallel transports along any transversal.

*Prove this using any assumptions you find necessary. Make as few assumptions as you can, and make them as simple as possible. **Be sure to state your assumptions clearly.***

What part of your proof does not work on a sphere?

To be able to prove Problem 20, as well as some other properties of parallel lines on the plane, it is necessary to make additional assumptions which have not been needed up to now. The conventional parallel postulates are Euclid's Fifth (and last) Postulate (EFP) and Playfair's Postulate (PP). Students may resist using EFP just as many mathematicians have, including Euclid himself. Over the centuries, mathematicians have made many other assumptions in order to avoid using EFP. Naturally, students will come up with many of these assumptions on their own. In this chapter, we will explore these various assumptions (those made by students and historical ones), their interrelationships, and their applicability to spheres. Students will then be in a position to make their own conclusions about each of the assumptions, deciding how they may be used in different contexts.

Therefore, our approach is to ask students to prove Problem 20 either by using any of the conventional postulates or by making their own postulate. Students will generally prefer to make their own parallel postulate than to use EFP or PP, which is one of the reasons for the approach used in this course. Ask students to make the following steps in their proofs:

· State clearly any assumption(s) they use which they are not able to prove with the results obtained in the course up to this point;

· Make as few and as simple assumptions as they can; and,

· Begin to decide whether this theorem is true on a sphere. If not true, ask them to explain which part of their proof on the plane does not work for a sphere.

A common argument used by students to prove Problem 20 on the plane is the following: "The midpoint of the transversal *t* can be extended in both directions to a line that is midway between the two lines. So, any line *s* that is a transversal will cross that midway line. At that point of intersection there must be the same symmetry properties as at the midpoint of *t*. Thus *s* must be a line of parallel transport also." This construction serves as a very useful example on a sphere. One can construct the same argument on a sphere, and thus if this argument were correct then the result would also be true on a sphere. Yet, this result contradicts what was proven in Problems 16 and 17. In fact, the angle situation on a sphere is very different for every point of the line that lies midway between *l* and *l'*.

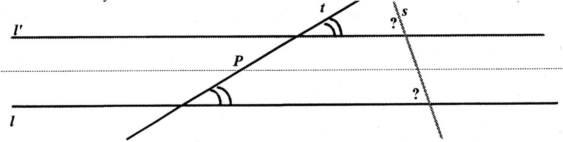

The line of "midpoints"

We will now state some parallel postulates or students' postulates to exemplify, once more, that the learning-teaching approach of this course brings out students diverse and creative ideas experiencing geometry. We emphasize that the assumptions or postulates should not be given to the students; rather it is crucial that the students construct their own postulates. You should make it a goal of your teaching to bring out students' ideas.

ASSUMPTION 1:

A quadrilateral with three right angles is a rectangle (i.e. all four angles are right angles).

In the figure below, *t* cuts *l* and *l'* at congruent angles, and *r* is also transversal to *l* and *l'*. Let *P* be the midpoint of the segment of *t* between *l* and *l'*. Consider a line *t'* which is transversal to l and l', perpendicular to l, and going through P. By the proof of Problem 17, t' also cuts l' perpendicularly. Parallel transport t' along l in such a way that the transported line r' goes through the midpoint, Q, of the segment of r between l and l'. The assumption implies that this action constructs a rectangle, $AA'CC'$. However, if that is the case, then the transversal r' cuts l and l' at congruent angles, and, by SAA, the midpoint for r is the same as the midpoint for r'. Thus (by Problem 17), since l and l' are parallel transports along r' then they also must be parallel transports along r.

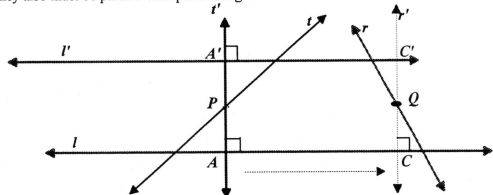

ASSUMPTION 2:

> *Parallel transported lines are equidistant.*

If l is a parallel transport of l' along t, then draw the line r perpendicular to l and l' through the midpoint of t. Now parallel transport t and r along l to any t' and r'. By the assumption, the portions of t and r that lie between l and l' are congruent to the portions of t' and r' that lie between l and l'. Then by SAS the two small triangles formed by t' and r' are congruent and thus l and l' are parallel transports along both t' and r'. Since r' can be placed anywhere, we can easily see that l and l' are parallel transports along any transversal.

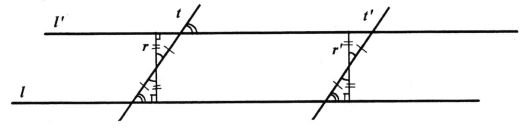

ASSUMPTION 3:

> *If two triangles have two corresponding angles congruent then the third angle is also congruent.*

> ***Proof without words:***

The following two assumptions can use this same proof.

ASSUMPTION 4:

> *The sum of the angles of a triangle is constant.*

ASSUMPTION 5:

> *The sum of the angles of a triangle is 180°.*
>
> (We will examine this assumption in Problem 23.)

ASSUMPTION 6:

> *Every point on "the centerline" is a center of half-turn symmetry.*

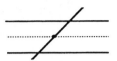

ASSUMPTION 7:

> *If α is not congruent to β, then the lines intersect.*

ASSUMPTION 8:

> *A transversal can be translated.* (Or, *parallelograms exist.*)

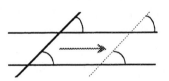

EUCLID'S FIFTH POSTULATE (EFP):

> *Pictorially:*

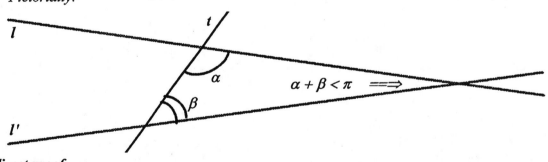

Indirect proof:

This proof uses the contrapositive version of EFP. Namely, *if two lines do not intersect then the sum of the interior angles formed by the two lines and a given transversal is greater than or equal to π.* The fact that it is not possible to use EFP directly might be one reason why students avoid using it.

Since the lines *l* and *l′* are parallel transports of each other, it follows from Problem 16 that they do not intersect. In the case depicted by the figure below, the contrapositive of EFP says that

$\alpha + \beta \geq \pi$ and $\pi - \alpha + (\pi - \beta) \geq \pi$;   that is, $\alpha + \beta \geq \pi$ and $\alpha + \beta \leq \pi$.   Consequently, $\alpha + \beta = \pi$, which means that lines l and l' are parallel transports along s.

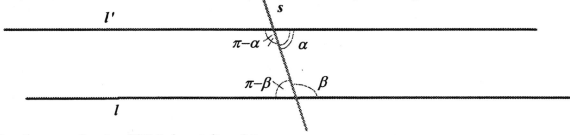

Another proof, *using EFP "almost directly"*:

Let α be the angle between l and r. Parallel transport l along r to B. If α is not the angle between l' and r, there are two possible cases. These are expressed in the figure below. In Case One we have that $(\pi - \alpha) + \beta > \pi$; in Case Two, $\alpha + \gamma > \alpha + (\pi - \alpha) = \pi$. In either of these cases, it follows from EFP that the lines l and l' intersect, but this is not possible since on the plane parallel transported lines do not intersect.

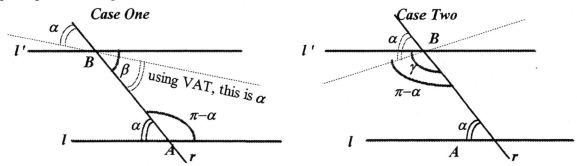

PLAYFAIR'S POSTULATE:

> *Given a line l and a point P not on l, there is a unique line r that passes through P and does not intersect l.*

Given the two parallel transported lines l and l' and another transversal s, parallel transport l along s to P the intersection of s and l'. By Problem 16, this transported line does not intersect l. By the assumption there is only one line through P that does not intersect l, that line must be l'.

Example of Students' Work on Problem 20

Professor and T.A. comments:

Student answer:

Problem 20 - Justin Collins

I want to first prove that, on a plane, if you are given two lines and a transversal that cuts them at congruent angles, then all transversals cut them at congruent angles. Consider:

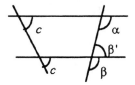

Assume we are given a line, cutting congruent angles *c*. For a second transversal that cuts angles alpha and beta, we need to prove that angle alpha is congruent to angle beta. Let's assume that α and β are not congruent. If they were congruent, then angle α and angle β′ would be supplementary. If α and β are not congruent, however, then α and β′ cannot be supplementary. They add up to either less that 180°, or more than 180°. Euclid's Fifth Postulate says that if these angles add up to less than two right angles, then lines one and two intersect on one side or the other. But this cannot happen because of Euclid's Exterior angle theorem.

If α and β′ are less than 180°, then the two lines intersect on the same side as these two angles:

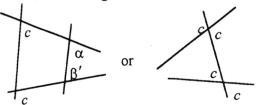

Good!

But this is forbidden by Euclid's Exterior angle theorem, because the exterior angle is equal to one of the interior angles (*c*). If α and β′ are greater than 180°, then the lines one and two intersect on the other side, but this is also forbidden for the same reason. (See above). Therefore, the assumption that angle α does not equal angle β must be false, and they must be congruent. If one transversal cuts two lines at congruent angles, then all transversals cut the lines at congruent angles.

On a sphere, this proof does not work. If we are given a transversal that cuts two lines at congruent angles, then we know that transversal must pass through the center. If we are given another transversal that does not pass through the midpoint of the first transversal, i.e. the center of the lune, then we know that it cannot cut at congruent angles because it must pass through the center. EEAT does not work universally on a sphere either. Our assumption on the plane that because the exterior angle is equal to the interior, there cannot be a triangle could also be wrong, particularly if lengths of 1/4 of a great circle are involved.

PROBLEM 21. *Parallel Postulates on the Plane*

On the plane, are EFP, Playfair's Postulate, and your postulate from Problem 20 equivalent? Why? or why not?

Most students have difficulty completing Problem 21, but we find that the effort is worth it in terms of the increased understanding they gain of the postulates.

Example of Students' Work for Problem 21

Professor and T.A. comments: *Student answer:*

Frank Sinnock - Problem 21

Let's prove Playfair's Postulate using EFP. To start, we have l and P.

Now, it must be shown there is only one parallel line through P.

Draw l' such that...? Imagine we draw t which passes through P and intersects l. Now, let's try to draw l'.

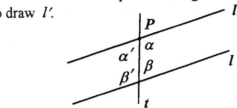

From EFP, though, we know if $\alpha + \beta < 180°$, then l intersects l'. If $\alpha + \beta > 180°$, $\alpha' + \beta' < 180°$ and l intersects l' again. Only if $\alpha + \beta = 180°$ does l' not intersect l, because $\beta \cong \alpha'$ and we have a transversal line like in Problem 16.

Why is l' unique? Now, let's prove EFP from Playfair's Postulate.

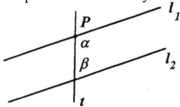

There is only one correct l_1 which is parallel to l_2 and goes through point P (Playfair's Postulate). From 16, we know this only happens when $180° - \alpha = \beta$ or $\alpha + \beta = 180°$. Therefore, if $\alpha + \beta \neq 180°$, it must intersect l_2.

My postulate is:

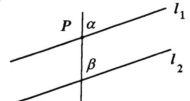

If α not congruent to β, then l_1 intersects l_2.

There is only one instance (namely $\alpha \cong \beta$) when l_1 is parallel to l_2, so this could be said to prove Playfair's Postulate.

The reverse is pretty much the same argument as proving EFP assuming Playfair's Postulate. We again use knowledge of 16 in conjunction with Playfair's to prove my postulate. My postulate is similar to EFP, but not as strong.

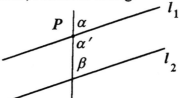

Given α and β, prove my postulate using EFP. $\alpha' = 180° - \alpha$. EFP says if $\alpha' + \beta < 180°$, then intersect on that side. If $\alpha' + \beta > 180°$, they intersect on the other side. If $\alpha' + \beta = 180°$, l_1 and l_2 are parallel.

Substituting for α': $180° - \alpha + \beta \iff 180°$.
And we find: If $\alpha > \beta$, l_1 intersects l_2. If $\alpha < \beta$, l_1 intersects l_2. And only if $\alpha = \beta$ does l_1 become parallel to l_2. Therefore, EFP proves my postulate.

Using the same diagram, we know that if $\alpha \neq \beta$, l_1 intersects l_2. Substituting for α, we find $180° - \alpha' \neq \beta$, $\alpha' = \beta \neq 180°$.

Then l_1 intersects l_2. Using symmetry or a triangle rule will indicate which side the lines intersect on, thus completing the proof of EFP using my postulate.

For many of the proofs, it was necessary to use information from another theorem. Theorems are not "equivalent," but they are very similar. Mine is actually a weak version of EFP (mine doesn't mention which side they meet on). Playfair's Postulate and EFP say similar things in different ways.

Response to comments:
In proving Playfair's Postulate using EFP, you inquired as to why there is only one unique line l' that passed through P and was parallel to l. From the proof in the last assignment, we know that for a transversal t

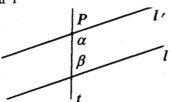

l' must form α such that $\alpha + \beta = 180°$. Therefore, there is only one possible l' for that transversal. There could be a different l' for each transversal with different values of β (t must always pass through P). But, from 20 we know that if one transversal has a parallel transport, all transversals through them have them as parallel transports. Since there can be only one l' for each transversal, they must all be the same. The l' through P is unique.

Are you using EFP and 20 to prove Playfair's Postulate?

Response to comments:

Last time you objected to my use of 20 to prove Playfair's Postulate using EFP. So, here we go again:

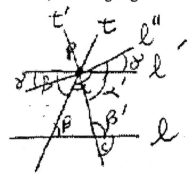

The problem is to prove that l', parallel to l, is unique. From EFP, we know that the only time l does not intersect l' with a transversal t when $\alpha + \beta = 180°$. But then, how do we know l' is the same for each transversal?

t' is another transversal that makes l'' parallel to l. l'' and l' do not intersect l, but are different. Let us say angle γ is between them as shown above. If that is so, then the situation of $l'', l',$ and t shows:

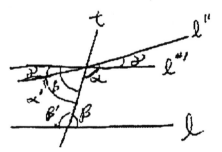

The new α' and β' are $\alpha' = \beta - \gamma$, $\beta' = \alpha$,

Where are you using EFP?

$\alpha' + \beta' = \alpha + \beta - \gamma = 180°$ which isn't possible unless γ is 0° and, therefore, l' coincides with l''. Therefore, there is a unique l'.

Response to comments:

In the last homework, you asked about where EFP was used to prove Playfair's postulate. In the last homework, right after the first picture, I wrote "From EFP, we know that..." This is because, back in my first answer to this question, I proved the idea that followed and did not wish to repeat myself. The first homework was just going on to prove that the line parallel to l through P was unique give what was already known from the homework before.

PROBLEM 22. *Parallel Postulates on a Sphere*

In many conventional geometry books, we can find the claim that EFP and PP are equivalent. Yet in Problem 21, we saw that to prove PP on the plane, EFP as well as the fact that two lines intersect in at most one point are both needed. Now we will see what happens on a sphere.

Starting with Problem 22(a), the students will prove EFP in a *strong sense* on a sphere. We say strong sense because EFP is trivially true on a sphere. In part (b) they will prove that PP is not true on a sphere, showing that PP and EFP cannot be considered globally equivalent. Finally, in part (c), students will relate their own parallel postulate and PP to a sphere by restating the postulates so that they work on a sphere. This reinstatement will involve parallel transport. Students may want to engage in a discussion about parallelism, parallel transport and the surfaces we have been working on, that is, the plane, sphere, and cones with cone angle bigger than 360°. Incorporating the hyperbolic plane, as closely related to a cone with cone angle bigger than 360°, seems appropriate at this point.

EFP on a Sphere in the "Strong Sense"

(a) Show that EFP is true on a sphere in a strong sense; that is, if lines l and l' are cut by a transversal t such that the sum of the interior angles $\alpha + \beta$ on one side is less than two right angles then, not only do l and l' intersect but, they intersect "closest" to t on the side of α and β. You will have to determine an appropriate meaning for "closest."

Let us look at some of the approaches students used:

APPROACH 1.

If $\alpha + \beta$ is less than two right angles then they are interior angles of the triangle formed by l, l', and t which does not contain the center point of the lune formed by l and l'.

Proof:

Assume the triangle that contains α and β in its interior also contains the center point of the lune. Consider the geometric figure resulting when the lune and transversal are half-turned around the center point (this can be done because the lune has half-turn symmetry around its center; see figure below). Then, because the triangle contains α and β does contain the center of the lune, the quadrilateral $ABB'A'$ is formed. *Since the quadrilateral can be divided into two triangles by a diagonal, we can conclude from Problem 23 that the sum of the angles of the quadrilateral $(\alpha + \beta + \alpha + \beta)$ is more than 360° and thus that $\alpha + \beta > 180°$.* If $\alpha + \beta$ were less than 180° then the triangle could not contain the center of the lune.

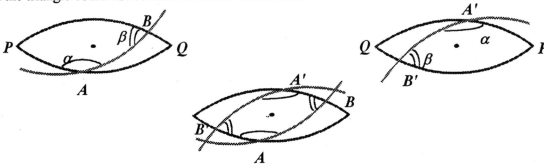

APPROACH 2.

If $\alpha + \beta$ is less than two right angles, then α and β are interior angles of the triangle formed by the transversal and the lune which has a median that is less than 1/4 great circle.

Proof:

Consider a transversal to a lune. If the transversal goes through the center of the lune, according to what was proven in Problems 16 and 17, the transversal cuts at congruent angles. We have shown that $\alpha + \beta$ are equal to a straight angle. Now consider the case in which the transversal does not go through the center of the lune. The triangle ABQ, in the figure below, has α and β as interior angles. If $\alpha + \beta$ is less than two right angles, then $\gamma + \alpha = \pi$, and thus, $\gamma > \beta$. It was shown earlier, in our discussion of EEAT, that in order for $\gamma > \beta$ to hold, the median, m, of AB must be less than 1/4 great circle.

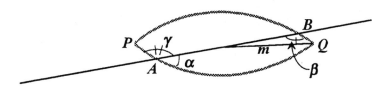

APPROACH 3.

If $\alpha + \beta$ is less than two right angles then the triangle containing α and β half-turns inside of the lune. If $\alpha + \beta$ is larger than two right angles then they are part of a triangle which half-turns outside of the lune.

Proof:

Note: From the work done on Problems 16 and 17, we can conclude that the triangles formed by intersecting a lune with a transversal are not congruent since one of the triangles fits inside of the other when half-turned around the center of the transversal. Half-turn one of the triangles around the center of the transversal. Two cases are possible. If α and β are part of the triangle that is inside the lune after half-turning, then the sum of α and β is going to be less than the straight line that goes through P, B and Q (see figure below, Case One). If α and β are part of the triangle that is outside the lune after half-turning, then their sum is going to be greater than the straight line that goes through P, B and Q (see figure below, Case Two).

Case One   **Case Two**

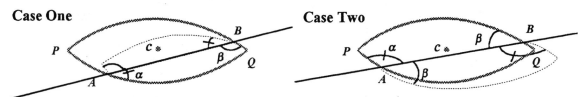

Thus, one can prove Euclid's Fifth Postulate in a strong sense on a sphere. You may want to point out that when one thinks about geometry in an intrinsic local way on a sphere, one finds a geometry that does not need the Fifth Postulate. One does not need to assume the Fifth Postulate on a sphere because it can in fact be proven!

Example of Students' Work on Problem 22

Student answer:

Problem 22 - Debra Van Savage

On a sphere, EFP is true in a weak sense because any two lines will intersect twice, on both sides of the transversal.

However, it is also true in a strong sense: l_1 and l_2 will intersect "closest" to l on the side of l on which are α and β.

If we look at the different cases:

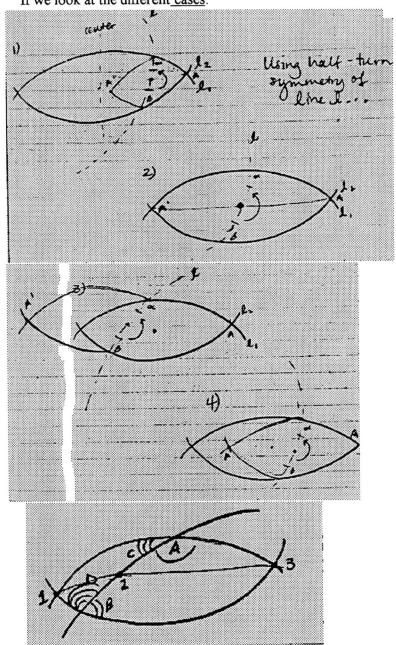

I'm not sure if I have this right:

If a transversal cuts two geodesics, on one side of the transversal, the two inside angles $(A)+(B)$ will have a sum that is greater than π, and on the other side, the two (interior) inside angles $(C)+(D)$ will have a sum that is less that π. (Assuming the transversal doesn't go through the center of the lune.)

I have to show that the distance from points 1 to 2 (the side that has $(C)+(D)$ as the interior angles) is <u>SHORTER</u> than the distance between 2 and 3 (the side with $(A)+(B)$ as its interior angles).

From the last page:
If we take the triangle we are working with to always be the <u>right</u> triangle and α and β to be interior angles of that triangle, (the triangle on the right side of the lune), we can see a pattern.

In 1) and 4), the transversal cuts on the right half of the center of the lune and the triangle is "small" (less than half a lune large) and when half-turn symmetry is used on line l, A' will always lie <u>inside</u> the lune.

However, in 3) the transversal cuts on the left half of the center of the lune and the triangle is "large" (greater than half a lune large) and when half-turn symmetry is used on line l, A' will fall <u>outside</u> of the lune.

In 2) the transversal cuts through the center of the lune, and therefore, $\alpha+\beta = \pi$

Here in this picture, if we keep parallel transporting l, the sum of α and β is less than π before l reaches the center. When l reaches the center, the sum is equal to π, and when l passes the center, the sum of will be greater than π.

To sum up, an appropriate meaning for "closest" would be [when α and β are the interior angles of the triangle formed by the transversal and the other two geodesics].

Where A' falls <u>inside</u> of the lune when half-turn symmetry is applied on line l.

Good

When A' falls <u>on</u> or <u>outside</u> of the lune after half-turn symmetry, the sum of the angles is either equal to or greater than π, meaning that the distance represented is <u>equal</u> to or <u>greater than</u> 1/4 great circle, and the intersection is <u>not</u> the closest.

Modifying Playfair's Postulate to Work on a Sphere

(b) *Using the notion of parallel transport, change Playfair's Postulate so that it is true on a sphere. Make as few alterations as possible and keep some form of uniqueness.*

Playfair's Postulate is totally false on a sphere because non-intersecting lines do not exist. If, however, we change "parallel" in Playfair's Postulate to "parallel transport", then <u>every</u> great circle through P is a parallel transport of l along some transversal.

Let us look at ways students have modified Playfair's Postulate so that it is true on a sphere:

· *For every line l and point P there exists a unique parallel transport of l along any transversal through P.*

· *For every line l and line $s \perp l$ and point P not on l, there exists a unique line l' such that s is an equator for the intersections of l and l'.*

· *For every geodesic, l and point P not on l and not a pole of l, there exists a unique line l' such that P is the center of the lune formed by those geodesics. Then every transversal going through P cuts l and l' at congruent angles.*

· *For every line l and point P there are infinitely many lines such that for every one of those lines l', there is only one transversal t that cuts l and l' at congruent angles and goes through P.*

Other Parallel Postulates on a Sphere

(c) *Either prove your postulate from Problem 20 on a sphere or change it, with as few alterations as possible, so that it is true on a sphere.*

We will now analyze the assumptions that the students used to solve Problem 20 on a sphere. We will restate them, where possible, in order to obtain a "parallel transport postulate" on a sphere!

ASSUMPTION 1:

If three angles of a quadrilateral are right angles, then the fourth is also.

This is false on a sphere (to see this, draw a diagonal and use Problem 23). It can be modified in the following way:

If three angles of a quadrilateral are right angles, then the fourth angle is larger than a right angle.

or,

If the two top angles of a quadrilateral on a sphere are right angles and the two adjacent sides are congruent, then the two base angles are congruent.

This last statement will be proven later in Chapter 12 where we will call a quadrilateral constructed in this way a *Khayyam Quadrilateral*.

ASSUMPTION 2:

Parallel transported lines are equidistant.

This is clearly false on a sphere since all great circles on a sphere intersect. There does not seem to be a straightforward way to change this statement to make it hold on a sphere. Notice that the half-turn symmetry property from Problem 16 is related.

None of the three assumptions that follow is true on a sphere:

ASSUMPTION 3:

If two triangles have two corresponding angles congruent, then the third angle is also congruent.

ASSUMPTION 4:

The sum of the angles of a triangle is constant.

ASSUMPTION 5:

The sum of the angles of a triangle is 180°.

The sum of the angles of a triangle on a sphere is always larger than 180° (students will prove this in Problem 23). AAA (Problem 19) is a related theorem that is true on the sphere. In addition, it follows from Problem 8 that:

Two triangles with equal area have equal angle sum.

ASSUMPTION 6:

Every point on "the centerline" is a center of half-turn symmetry.

On a sphere only one point on the centerline is a center of half-turn symmetry: the center of the lune made by the two lines.

ASSUMPTION 7:

If α is not congruent to β, then the lines intersect.

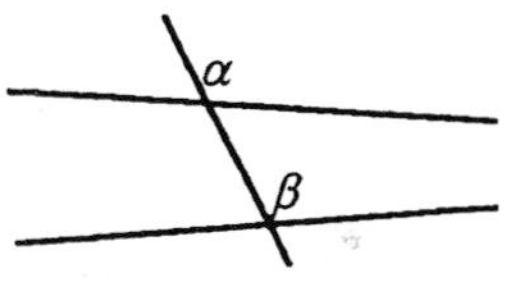

ASSUMPTION 8:

If α is not congruent to β, then the lines converge.

Assumptions 7 and 8 do not make sense on a sphere given that lines always converge and intersect on a sphere, whether or not the transversal cuts at congruent angles.

ASSUMPTION 9:

A transversal can be translated. (Or, *parallelograms exists.*)

On a sphere translating the transversal along a lune will change the angles since a transversal only cuts at congruent angles if it goes through the center of the lune. In Chapter 12 we will investigate *Khayyam Parallelograms* which are analogs on a sphere of planar parallelograms.

PROBLEM 23. Sum of the Angles of a Triangle

What is the sum of the angles of a triangle on the plane and on a sphere?

Some students may have used Problem 23 in their solutions to other problems. They should be reminded that they cannot use any of those results in Problem 23.

The sum of the angles of a triangle on the plane is equal to a straight angle.

Proof, pictorially:

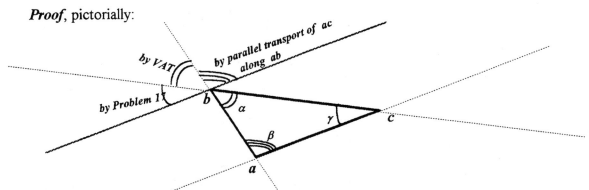

Students know that on a sphere the sum of the angles of a triangle can be bigger than two right angles. For example, a triangle on a sphere can have three right angles. The question then is:

Do the angles of a triangle on a sphere ever sum to two right angles or less?

You might ask students to prove that the sum is always more than two right angles, or ask them to find a triangle on a sphere for which the sum of the angles is less than two right angles and prove that it is so. Some students will propose that the sum of the interior angles of a triangle is a straight angle if the triangles are small enough to be seen as planar or infinitesimally planar. This is valid for infinitesimal triangles, but we are interested primarily in spherical triangles that are not infinitesimal. Let us look at a proof of the following result:

The sum of the angles of a triangle on a sphere is always larger than a straight angle.

Proof:

Consider the side AB of $\triangle ABC$. The geodesic that contains AB cuts the lune that contains the segments CB and CA. If this triangle contains the center of the lune, then its median is bigger than 1/4 great circle, and $\alpha + \beta \geq 180°$ from which it follows that $\alpha + \beta + \angle BCA > 180°$. If $\triangle ABC$ does not contain the center of the lune of CB-CA, then *parallel transport $\angle BAC$ along AC to C.* The transport and the line AB form a lune for which AC's midpoint is the center. Now, using the midpoint of BC, apply a half-turn rotation transformation to $\triangle ABC$. It follows from EEAT (Problem 15) that the exterior angles are larger than the interior angles, and thus, that the rotated triangle lies as pictured below.

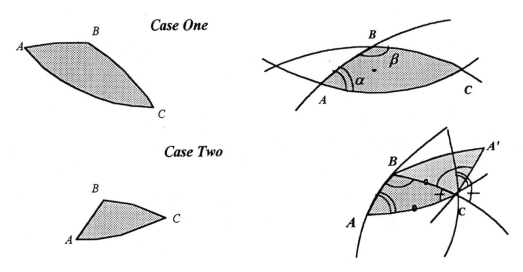

Proof using parallel latitudes:

Here is a proof discovered by a freshman English major who was taking a course for "students who did not yet feel comfortable with mathematics."

First note that:

Two latitude circles which are symmetric about the equator have the property that every transversal has opposite interior angles congruent.

This follows because the two latitudes have half-turn symmetry about any point on the equator.

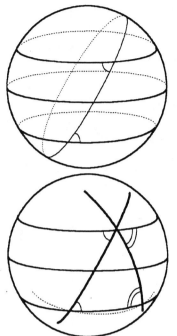

Now we can mimic the planar proof:

We see that the sum of the angles of the "triangle" in the figure sum to a straight angle. This is not a true spherical triangle because the base is a segment of a latitude circle instead of a great circle. If we replace this latitude segment by a great circle segment then the base angles will increase. Clearly then, the angles of the resulting spherical triangle sum to more than a straight angle.

You can check that any small spherical triangle can be derived in this manner.

Example of Students' Work on Problem 23

Professor and T.A. comments: *Student answer:*

Problem 23 - Sylvan Kavanangh

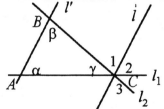

Given any $\triangle ABC$, construct L parallel to AB through C.

Since we proved that any transversal cuts parallel lines at congruent angles (#21), we know that $\angle 3 \cong \beta$ and $\angle 2 \cong \alpha$ (l_1 and l_2 are transversals). Now, $\angle 1 \cong \angle 3 \cong \beta$ (opp angles $\cong$)

Now we have this configuration:

and then $\gamma + \beta + \alpha =$ angle in a straight line.

<u>Sphere</u>:

I will draw the same configuration:

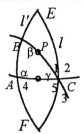

When I construct l , I can make a parallel transport of α or β . I chose β .

Because of the way I made this construction, $\angle 3 \cong \beta$ and so $\angle 1 \cong \beta$ (opp angles $\cong$)

$\angle 2$ is not congruent to α because P , is not equidistant from E and F . Not a parallel transport. So, $\alpha + \beta + \gamma$ is not congruent to π .

Furthermore, in this case, I can say that:

$\angle 4 + \angle 5 < \pi$ (EFP), $\angle 4 = \pi - \alpha$ (straight line),

and then $\pi - \alpha + \angle 5 < \pi$, but then: $\angle 5 - \alpha < 0$; $\angle 5 < \alpha$; $\angle 2 < \alpha$.

Since $\alpha > \angle 2$ and $\gamma + \beta + \angle 2 \cong \pi$, $\gamma + \alpha + \beta > \pi$.

Good.

Chapter 11

3-Spheres in 4-Space

In this chapter, students will use knowledge about two-dimensional spheres to explore, in both an intrinsic and an extrinsic way, a three-dimensional sphere in four-dimensional space.

It is difficult for many people to imagine four dimensional space, so we begin by encouraging students to build an image of what 4-space might be like. We do this by having them explain 3-space to a 2-dimensional person.

PROBLEM 24. Explain 3-Space to 2-Dimensional Person

How would you explain 3-space to a person living in two dimensions?

Think about the question in terms of this example: The person below lives in a 2-dimensional plane. The person is wearing a mitten on the right hand. Notice that there is no front or back side to the mitten for the 2-D person. The mitten is just a thick line around her hand.

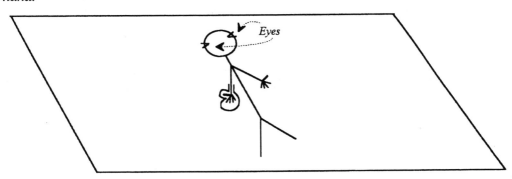

Suppose that you approach the plane, remove the mitten, and put it on the 2-D person's left hand. There's no way within 2-space to move the mitten to fit the other hand. So, you take the mitten off of the 2-D plane, flip it over in 3-space, and then put it back on the plane around the left hand. The 2-D person has no experience of 3 dimensions but can see the result — the mitten disappears from the right hand, the mitten is gone for a moment, and then it is on the left hand.

The mitten is gone
for a moment

and then...

How would you explain to the 2-D person what happened to the mitten?

The example of explaining to a 2-D person how a right-handed mitten can travel through 3-space and become left-handed, provides a concrete example in the 2-D person's world that involves a discussion of 3-space. At first, students find the task of trying to explain to the 2-D person where the mitten went very difficult and somewhat confusing. After all, many students say, if the 2-D person has no experience of 3-space, how will the person ever be able to understand an explanation of it? We try to stress the idea that it *is* possible to have *images* of things we cannot directly experience (for example, we have an image of an entire sphere inside and out), and it is possible for the 2-D person to do the same.

Note that a 2-D person living in a plane perceives only the plane. Their plane of existence extends indefinitely to their left, right, top, and bottom — but they lack any sense of front and back. In their world, mittens go around their hands like a thick line and doesn't cover the hand from our 3-D point of view:

It is important to encourage students to explore the images and experiences they have of different dimensions. Comparing one's own images and experiences with those of the two dimensional person is helpful for building images of the 4th dimension.

Examples of Students' Work on Problem 24

Professor and T.A. comments: *Student's answers*

Problem 24 - James Kessler

Since the 2-dimensional person understands and visualizes a 2-dimensional world, it is in these terms that 3-space must be explained. The person knows about planes, lines and points, therefore, we can attempt to use these to help the 2-D person conceptualize our dimension. To begin with we can explain the idea of perpendicular (and other general angles) formed by the intersection of two lines by having them look at their fingers — e.g. the alignment of the thumb and first finger are perpendicular when the fingers are stretched. Now it can be explained that a line in their 2-dimensional universe is caused by the intersection of another two dimensional plane that acts much like the intersection of the lines of her fingers. It might be slightly easier for the person to visualize this if they are asked to imagine only the edges of the two planes (i.e. straight lines), thereby reducing the concept to a figure much like their thumb and finger.

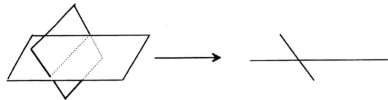

Nice explanation!

Then it can be easily explained that her mitten was transported on another 2-D plane as it rotated through the line of her body until the planes coincided once again. Again it may help to have the 2-D person think in terms of lines and segments, so it could be explained that the movement is similar to her moving the line segment (what she can see) of the mitten in a half-circle on her own plane.

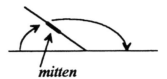

mitten

These explanations should help the 2-D person to begin to understand and to visualize 3-space, if not experience it.

Problem 24 - Laura Kozuh

In trying to explain to a 2-dimensional person how his right mitten can be flipped over to become his left mitten, I have become extremely confused. Since he can only see things in two-space, how can you just flip it over, which would involve 3-space?.

I (or you) am (are) doing the flipping, and then explaining to him what we did.

What is this 4th dimension like for you? Perhaps explaining this to me, who can't experience it will help you in finding a way to explain to the man what happened to his mittens.

*Good. You are using something that the man **can** experience to talk about the dimensions. Use this to explain to him about the mitten. This is closely related to taking an extrinsic vs an intrinsic point of view.*

The only idea I have really come up with so far is that you could take the plane he is on and fold it along his middle. The plane has not gone away, it's just moving. However, for it to do so it must make use of 3-space, which the person has never experienced. But just like we have a four-dimensional concept of an object in our minds even though we are only in 3-space, perhaps the 2-d person could have a concept of an object in 3-space. In this way, he might be able to understand the movement of the plane he is in. He will still be able to see what is on the plane as usual, and he will remain on the plane, but his plane of existence is moving in three-space.

For example, if you have a line in one-space the only way that the [gray] dots and the [white] dots can match up is for the line to be bent in the middle and for it to enter 2-space.

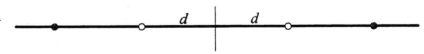

They cannot just slide along the line. So someone on that line would have to have a concept of 2 space to understand what was happening.

Unfortunately, I am extremely confused by all of this and am not sure how or where to begin! I understand how to flip the mitten, but I don't know how to explain it to the 2-dimensional person.

Response to comments:

I guess the fourth dimension to me would be like sitting in my room closing my eyes, and being able to imagine every aspect of the room all at once. In three dimensions, it is impossible, without looking from above, to see the entire room. You will always be missing part of it. But in four dimensions, you could see every wall and every object simultaneously. One way I thought of to picture it is like a movie. In movies we are able to visualize three-dimensional things on a plane. In a way, you could consider being in 4-D sort of like watching a 3-D movie but from every angle all at once.

nice.

Likewise, I figured that one way to explain 3-D to the man in 2-D would be to set him in the middle of a circle and have him look around and try to imagine being able to see all of it at once, just like trying to visualize your own room. To be able to do this would be to have an understanding of these dimensions, because only in 3-D could you see the entire circle at once. Another thing I thought of but that I'm not quite sure how to apply, is a pop-up book. All the little objects can be folded into a plane but when you open them up, they pop into three-space. It would be useful for him to understand this concept but I'm not quite sure how to explain it to him!

Also, if the little man can see in 2-D, intrinsically, how the dots on the 1-D line must come together in order to match up (shown in the drawing) and how they must enter 2-space to do so, then we can explain to him that this is the same way his mitten can get moved to the other side. He will have to visualize it extrinsically and to imagine his plane being like the 1-D line and his mitten being like one of the dots on that line. In this way, he might be able to understand how his mitten must enter another dimension in order to match up on the other side.

excellent.

Visualizing 4-Space

Some students will try to draw pictures of projections in 3-space of objects in 4-space, e.g. the usual projections of the four dimensional hypercube. However, our experience has been that this does help the students to understand the geometry of 4-space. When we see the following drawings:

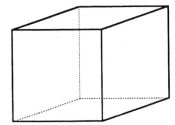

We recognize them as the 3-D shapes of a sphere and cube because we have had much experience with these 3-D shapes and their projections. These drawing help us to see the geometric properties of the 3-D shapes since we can imagine the whole shape. Since we have not had experience with 4-space objects projections, these objects cannot help us in the same way to understand the geometry of 4-space.

Instead of projections, the students should be encouraged to look at two- or three-dimensional cross-sections. Here the appropriate linear algebra fact is that any two [three] linearly independent vectors in *n*-space always determine a 2-[3-]dimensional subspace which is geometrically a plane [a 3-space]. Students should use linear algebra to whatever extent they are comfortable with it, but they should use it the way that facilitates their geometric understanding.

PROBLEM 25. *Intersecting Great Circles in the 3-Sphere*

Students are being asked to prove the following theorem:

If two great circles in S^3 intersect, then they lie on the same great 2-sphere.

Proof:

Let us assume that the two great circles do not coincide everywhere. Every plane that goes through the center of S^3 produces great circles that are centered on the center of S^3 and have radius R. But, given that all great circles (or great 2-spheres) have central symmetry relative to their common center, if the great circles intersect in one point, then, by symmetry, they also have to intersect at the opposite pole. Note that the planes that produce those great circles intersect in one line that contains the point-pair intersection of the great circles. Consider a vector from the center of S^3 to one of the intersection points, v. Consider, also, two other vectors that go from the center of S^3 to two other points which are neither of the antipodal points. One, v', is on one great circle the other, v'', on the other great circle. These three vectors are linearly independent and they generate a three dimensional subspace of R^4. Intersect this subspace with S^3 and you get, by definition, a great 2-sphere. This great 2-sphere contains both great circles, given that it contains three points of each great circle. Remember that a great circle is completely defined by two non-antipodal points or any three points.

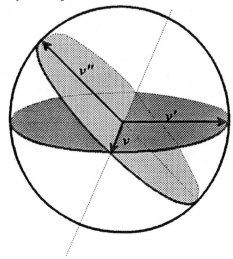

Professor and T.A.
comments:

Example of Students' Work on Problem 25

Student answer:

Problem 25 - Kevin Newman

A point in R^4 may be characterized by a 4-tuple such as (x, y, z, w) where each element represents a point in one of the dimensions X, Y, Z, and (arbitrarily named) W.

A great circle (in XYZ, say) is then characterized by $(x, y, z, 0)$ (where 0's represent non-points or points at the origin) as shown below:

This is a great sphere.

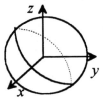

2 ?

A great 2-sphere is the sphere formed by any three of the four dimensions forming the 3-sphere. Any great circle resides in three dimensions; intersecting the four-dimensional sphere with a plane eliminates one of the dimensions since a plane of arbitrary orientation is defined by three dimensions. Now, if two great circles intersect, then both sets of three points defining the two circles must be equal at some point, which means they must be the same three dimensions or else they could never all specify the same point. If the three dimensions are the same, then a 2-sphere is formed by definition.

?

Response to comments:

When I was talking last time, if you look at the picture, you can see I was thinking of a great circle that tilted, i.e., not just in the XY plane, but in XY and with some linear component in Z. If you take the great circle to *define* two-dimensional subspace, then you are back down to two dimensions.

O.K.

If the two great circles intersect, then the two-dimensional subspaces which these great circles span are not linearly independent. Therefore, there is at least one basis vector which can be made up of the other three. If there were two such vectors, the great circles would not be distinct, therefore, there is only one. If we have, then, three independent basis vectors, then we have defined a three-dimensional subspace of the four we are working in. Hence a great 2-sphere is formed and the points in question must lie on it.

You could also use 26. Take two points from one of the great circles and one point from the other (not at the intersection) and if they're distinct, then a 2-sphere is defined.

PROBLEM 26. *Triangles in the 3-Sphere*

Show that if A, B, and C are three points in S³ that do not lie in the same great circle, then there is a unique great 2-sphere, G², containing A, B, and C.

Thus, we can define △ABC as the small triangle in G² with vertices A, B, C. With this definition, **triangles in S³ have all the properties of small spherical triangles which we have been studying.**

Proof:

Suppose that three points, A, B and C, do not all lie in the same great circle. We can choose at least two non-coinciding intersecting great circles that contain the three points. With the proof of the theorem in Problem 25 at our disposal, we can show that these two great circles lie on the same great 2-sphere. This great 2-sphere is unique because the three linearly independent vectors define a unique 3-dimensional subspace.

Note that the results of Problem 26 will allow us to define a small triangle on the surface of a great 2-sphere, and, hence, we have a definition for triangles in S³. By thinking of triangles in S³ as equivalent to triangles on great 2-spheres, we can easily convince ourselves that all the properties of spherical triangles that we have proved thus far will hold for triangles in S³.

Example of Students' Work on Problem 26

Student answer:

Problem 26 - Jeremy W. Schulman

Take the following three points in S³ that do not lie on the same great circle:

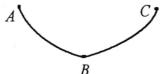

Say: A can connect to B and C can connect to B. Let A, B lie on the same g.c. and B, C lie on another g.c.

From 25, these two great circles that intersect lie on the same great two sphere. This great two sphere must be unique as a great two sphere is defined by two different great circles intersecting (hence 3 points on two separate great circles).

PROBLEM 27. *Disjoint Equidistant Great Circles*

Show that there are two great circles in S³ such that <u>every</u> point on one is a distance of one-fourth great circle away from <u>every</u> point on the other and vice versa.

Proof:

In R⁴ every 2-dimensional subspace has an *orthogonal complement* which is also a 2-dimensional subspace of R⁴. These 2-dimensional subspaces are planes that intersect only at the origin, and such that every line going through the origin on one of the subspaces is perpendicular to every line that goes through the origin on the orthogonal plane. Consider one great circle, *l*, which is defined by the intersection of a plane, Π, with S³. If we take the orthogonal plane Π′

and intersect that plane with S^3, we get a great circle: l', that is "orthogonal" to the great circle l. Take two points, P and P', one on each great circle, and consider the plane that goes through those points and the center, C, of the sphere. Call g the intersection of that plane with S^3. Since P and P' are on orthogonal planes, the segments PC and $P'C$ are orthogonal and thus the distance between the two points is 1/4 great circle.

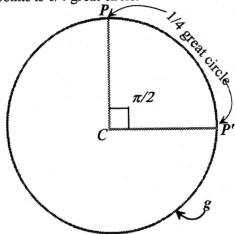

The distance (in S^3) from P to P' is 1/4 great circle or $\pi/2$.

Example of Students' Work on Problem 27

Student answer:

Problem 27 - Ted Eigel

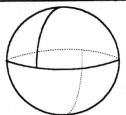

In S^2 I can show you a point pair that is 1/4 great circle from every point on a great circle.

So in S^3 the question is; are there 2 great circles that are 1/4 g.c. away from one another at every point? Yes!

There are two g.c.'s such that any point on either great circle is 1/4 great circle from the other. Although it is difficult to visualize, it is as if the point spatially "contained" another g.c. whose existence is not in the three dimensions that encompasses S^2. So the points are all as if they lay on the pole, as far as their relationship from the other sphere is concerned.

Many students see this by using coordinates in some way close to this: Consider the four co-ordinate directions in 4-space to be x, y, z, w. Then the 3-sphere is represented by:

$$x^2 + y^2 + z^2 + w^2 = R^2$$

and the two great circles are

$$x^2 + y^2 = R^2, \text{ in the } zw\text{-plane}$$

and

$$z^2 + w^2 = R^2, \text{ in the } xy\text{-plane.}$$

A Rotation That Moves All Points

The analogous situation on a 2-dimensional sphere occurs when you consider a pair of antipodal points and their corresponding equator. The two points lie in a line which is the orthogonal component of the plane in which the great circle lies. Consequently, the line is perpendicular to every line that goes through the center of the sphere on the orthogonal plane. And, each pole is a distance of 1/4 great circle from every point on the equator.

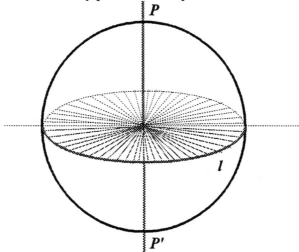

Looking at the picture above, we can see that a rotation of the 2-sphere along l will move all points of the sphere except the antipodal points P and P'. In the 3-sphere if we look at a great 2-sphere that contains the great circle l, we will see exactly the same picture except that now the antipodal points P and P' are on the great circle l' which is orthogonal to l. When we rotate S^3 along l, the whole of S^3 will move except for the orthogonal great circle l'. However, if we rotate along both l and l' simultaneously and at the same speed, then every point on S^3 will move. In fact, they will each move along a great circle.

Focusing on the diagram below, let us find an expression for this rotation. If we pick up a point with coordinates (x, y, z, w) on S^3, then $x^2 + y^2 + z^2 + w^2 = R^2$. The projections onto the xy-plane, zw-plane are:

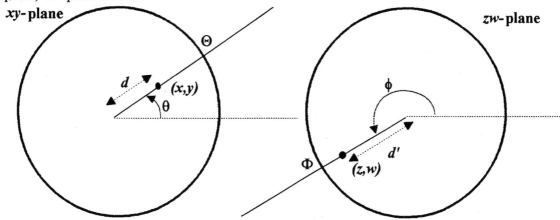

If we look at the plane determined by these two projections we get:

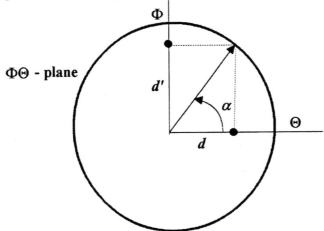

ΦΘ - plane

The coordinates for a point in S^3 can be given by (θ, ϕ, α), where $0 \leq \theta, \phi < 2\pi$ and $0 \leq \alpha \leq \pi/2$, and the rotation through angle τ takes point (θ, ϕ, α) to point $(\theta+\tau, \phi+\tau, \alpha)$.

It is not obvious to see, but nevertheless true that the path of each point P of S^3 during the rotation describes a great circle. To see this we first simplify the algebra by re-coordinating so that $P = (a, 0, b, 0)$, where $a^2 + b^2 = R^2$. Then the rotation through angle τ takes P to the point $(a\cos\tau, a\sin\tau, b\cos\tau, b\sin\tau) = \cos\tau\,(a, 0, b, 0) + \sin\tau\,(0, a, 0, b)$, which is the parametric equation of a circle with center at the origin and radius $a^2 + b^2 = R^2$ and is thus a great circle. During the rotation, two different points travel on great circles which either coincide or are equidistant. These great circles are traditionally called **Clifford Parallels** and they are said to be a **fibering** of S^3. An accessible exposition of Clifford Parallels can be found in [**C**: Penrose].

PROBLEM 28. Is Our Universe a 3-Sphere?

"What is the shape of our universe?" is still an open question. Some people have hypothesized that it is spherical. In Problem 28, students will try to decide how we might be able to determine if the universe is spherical. In this sense, each student must act like a bug that cannot travel far away but can see distant objects.

> *Given that it is impractical to measure the excess of triangles in our solar system, how might we get information as to whether the universe is Euclidean or spherical by only taking measurements of angles within our solar system but looking to the stars. [If you have trouble conceptualizing a 3-sphere, then you can do this problem for a very small bug on a 2-sphere who can see distant points (stars) on the 2-sphere, but who is restricted to staying inside its "solar system" which is so small that any triangle in it has excess too small to measure.]*

If our universe were a 3-sphere and we could see with our telescopes all the way around the universe (the length of a great circle), then we could easily tell that our universe is 3-sphere because we would be able to see ourselves at a distance no matter which direction we looked in.

If we could see a little more than half-way around the universe, then we should be able to look in two opposite directions and see the same star which is situated near the point which is antipodal to us.

If we could see more than one-quarter of the way around the universe then we might be able to see the same star from two divergent directions. (See the figure below.) If a star can be sighted such that angles α and β have a sum greater than π, then the triangle must spherical since its angle sum to more than π.

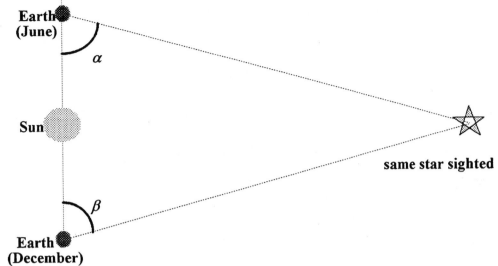

This method has practical problems: If the distance to the star was close to 1/4 great circle then the angles α and β would have a sum very near to π. Thus the angles would have to be measured with extreme accuracy because solar system distances are very much smaller than the distance to the star. For stars further away, $\alpha + \beta$ would become larger and thus the difference from π would more measurable.

The following student example gives another method based on the perceived distribution of the stars at various distances.

Example of Students' Work on Problem 28

Professor and T.A. comments:

Student answer:

Problem 28 - David Walend

Assume that stars are distributed evenly in the S^3 surface. (they aren't, but galaxies may be). If space is curved, lines will meet at a point in the distance (2-d projection of 3-d "biangles" in a 4-d space):

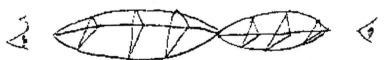

If stars are distributed evenly, more stars will be seen in the "antinodes" of the above picture than in the nodes. (Nothing will be seen at the nodes.) For long distance, star density would reach a maximum and then fade out. Fading off and may not be observable.

Also, for stars along on their great circle, each star would have a "long-way-round" component.

For great circles with no objects, the viewer would see him/herself.

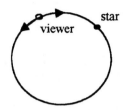

Explain more

    I like the possibility of a parabolic or hyperbolic space better, even though the shape makes the observer's position matter.

    This helps explain the 2.9 °K background without necessity of the big bang. Matter would only be able to get so far away no matter how fast it is going.

Chapter 12

Dissection Theory

As students work through Problems 29 to 37, they will confront relationships between geometry and the algebra of the real numbers, areas, similar triangles, the Pythagorean Theorem, and possibly the oldest written problem in geometry (possibly 3700 years old!). *Dissection theory* will be the tool for exploring the problems in this chapter and the next.

*Two geometric figures, F and G, are **equivalent by dissection** if one of the figures can be cut up into pieces and the pieces can be rearranged to form the other.*

We will refer to two geometric figures, F and G, that are equivalent by dissection as $F =_d G$.

*Two geometric figures, F and G, are **equivalent by subtraction** if one can find two other geometric figures, S and S', equivalent by dissection such that $F \cup S =_d G \cup S'$. F and S as well as G and S' intersect at most in their boundaries. We will write $F =_s G$.*

The notion of equivalent by subtraction is weaker than the notion of equivalent by dissection. All of the problems below can be approached using equivalence by dissection, but in some cases students will find equivalence by substraction easier to use. With these two definitions in mind, the students are ready to explore the meaning of area from a new perspective.

PROBLEM 29. *Dissecting Plane Triangles*

Show that on the plane every triangle is equivalent by dissection to a parallelogram with the same base no matter which base of the triangle you pick.

The proofs of Problem 29 are very similar to the proofs of the analogous result on the sphere (Problem 32). Thus we will discuss these proofs after Problem 32 has been introduced. However, it is best for the students to work on Problem 29 first. When discussing Problem 29 in class, many different approaches should emerge, and these will lead students to be better able to investigate Problem 32.

Example of Students' Work on Problem 29

Professor and T.A. comments:

Student answer:

<u>Problem 29 - Panfilos Luna</u>

Why is this a parallelogram? Proof?

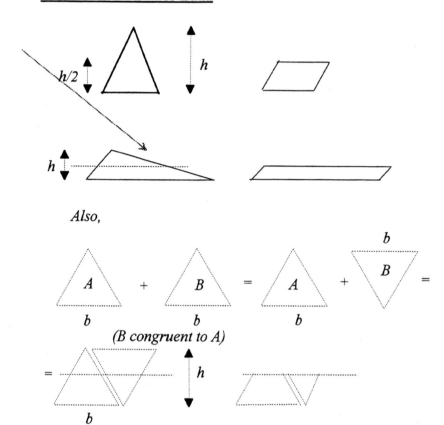

Also,

(B congruent to A)

Response to comments:

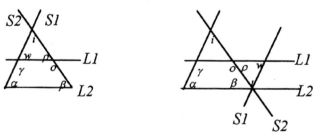

Why is 1). $AD \cong BC$

2). $AB \cong DC$ *and* AB *is parallel to* DC?

$L1$ is parallel to $L2$, $S1$ is a transversal of $L1$ and $L2$, so $\alpha \cong w$ and $\beta \cong \rho$, $o + \rho \cong \pi/2 + \pi/2$ ($S2$ rigid and straight)

$\alpha + \beta + i \cong \pi/2 + \pi/2 \cong \alpha + \gamma$; $\beta + i \cong \gamma$ Gives:

$L1$ parallel to $L2$

Response to comments:

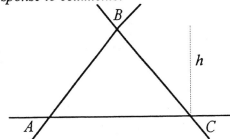

Draw a line parallel to AC at a height of $h/2$:

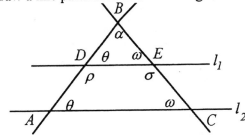

We know that $\angle BDE \cong \angle BAC$ and $\angle BED \cong \angle BCA$.
We also know that $\alpha + \theta + \omega = 180°$ and $\rho + \theta = \sigma + \omega = 180°$.
So $\gamma + \theta = \sigma$ and $\gamma + \omega = \rho$.

By rigidity, rotate $\triangle BED$ about point E until BE lies directly on EC

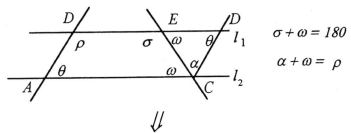

$\sigma + \omega = 180$

$\alpha + \omega = \rho$

$\Downarrow$

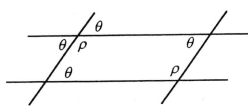

PROBLEM 30. Dissecting Parallelograms

Show that on a plane every parallelogram is equivalent by dissection to a rectangle with the same base and height.

As before, we will postpone discussing the proofs of Problem 30 until we examine the analogous problem on a sphere in Problem 33. Again, it is best for the students to work on and discuss Problem 30 before attempting Problem 33.

Example of Students' Work on Problem 30

Professor and T.A. comments:

Student answer:

Problem 30 - Linda Hermer

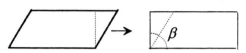

Every parallelogram is $=_d$ to a rectangle of same base and height.

Why is $\beta = 90°$?

Response to comments:

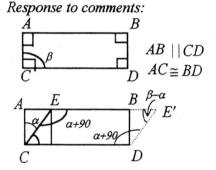

$AB \parallel CD$
$AC \cong BD$

Look at this case:

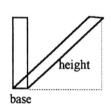

If at $\triangle ACE$ we cut and paste aligning AC with BD, then we have subtracted one congruent triangle and attached another (or the same one) at opposite end of rectangle. This forms a parallelogram with angles $\beta-\alpha$, $\beta+\alpha$, $\beta-\alpha$ and $\beta+\alpha$. It retains its height of AC (or BD) and its base of $AE + EB$.

Response to comments:

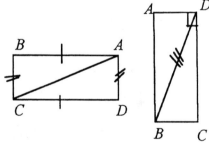

Rigidly separate $\triangle ADC$ and move so that DC is flush against AB:

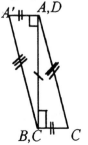

All angles unchanged through rigid move, except those divided into their components by triangle separation; angles still add to 2π.

Area of $ABCD$ = base · ht. (e.g. $BC{\cdot}AB$). Area of each parallelogram created by rigid movements of triangles still = area of two triangles together = $1/2 \cdot 2$ (base·ht.) = $BC{\cdot}AB$, the area of the original rectangle.

How do you go from parallelogram to rectangle in this case?

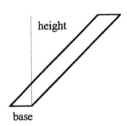

Response to comments:

You said "look at case:

".

If I start with a parallelogram, then vertically transect and parallel transport the triangles along L' to form other half of rectangle:

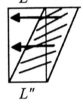

Yes, but:

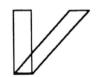

Response to Comments:
I'm going to give the easiest case to draw for a general solution to

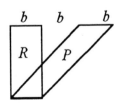

on the plane.

To go from the parallelogram to the rectangles by dissection (or vice-versa):

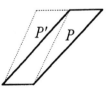

1) Go from P to P' by cutting out the right most triangle on P, parallel transporting it to the left far enough to complete P'.

2) Then, by the same procedure, go from P' to R. The procedure can be followed when P and R are separated by a non-integer multiple of b; e.g.:

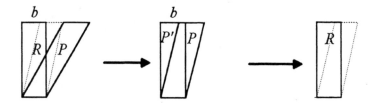

Dissection Theory on Spheres

The problems above take on a different form on a sphere, since one cannot construct a rectangle, *per se*, on a sphere. We define two kinds of polygons on a sphere and then restate the problems for a sphere. The two polygons are the Khayyam Quadrilateral and the Khayyam Parallelogram. These definitions were first put forth by Persian geometer-poet Omar al'Khayyam in the 11th century AD. Through a bit of Western Chauvinism, geometry books generally refer to these quadrilaterals as Saccheri Quadrilaterals after the Italian monk who translated and extended the works of al'Khayyam, and others, from Arabic.

A **Khayyam Quadrilateral** (KQ) is a four-sided polygon with two adjacent right angles, and with the adjacent sides to those angles congruent. In the figure below, $AB \cong CD$ and $\angle BAD \cong \angle ADC \cong \pi/2$. The right angles of the KQ are called *top angles*, BC is the *base* of the KQ, and the adjacent angles of the base are called *base angles*. One can check that a KQ on the plane is a rectangle.

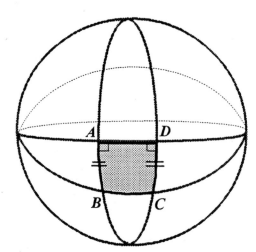

A **Khayyam Parallelogram**, KP, is a four-sided polygon such that two of its sides are congruent, and these same two sides are cut at congruent angles by another side. In the figure below, $AB \cong CD$ and AB is a parallel transport of DC along AD. BC is called the *base* of the KP and its adjacent angles are called the *base angles*. On the plane a KP is a parallelogram (to see this draw in a diagonal and use Problem 16).

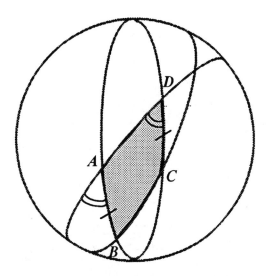

PROBLEM 31. *Khayyam Quadrilaterals*

Prove that the base angles of a Khayyam Quadrilateral are congruent and that the perpendicular bisector of the top is also the perpendicular bisector of the base.

Proof:

Connect points A and B to points C and D, respectively, and consider $\triangle ABD$ and $\triangle DCA$. Given that AD is a common side, $AB \cong DC$, and, by construction, $\angle BAD \cong \angle CDA$. Then, by SAS, one can conclude that $\triangle ABD \cong \triangle DCA$. But, $AC \cong BD$, so $\triangle ABC \cong \triangle DCB$. Thus, the base angles $\angle ABC$ and $\angle DCB$ are congruent.

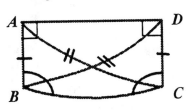 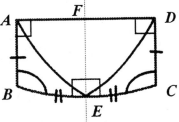

For the second part of the proof, E and F are the points of intersection of the perpendicular bisector of BC with AD, respectfully, the base and top of the KQ. Draw the segments AE and DE, and note that by SAS, $\triangle ABE \cong \triangle DCE$. But, by ASA, $\triangle AEF \cong \triangle DEF$. Therefore, $\angle AFE \cong \angle DFE$ and $\angle AFE + \angle DFE \cong \pi$. So, it follows that $\angle AFE \cong \angle DFE \cong \pi/2$. On the other hand, the congruency of the triangles also implies that $AF \cong DF$.

Another Proof:

Consider the perpendicular bisector of AD. Since $\angle EFA \cong \angle EFD$, $AF \cong DF$, $\angle FAB \cong \angle FDC$ and $AB \cong DC$, reflection through FE will take F, D, C to F, A, B. The segment EC will reflect to EB since they are the unique short great circle segments connecting the endpoints. Thus, the quadrilateral has reflection symmetry relative to this bisector. That is, the symmetry of the top also holds at the base. Thus, the base angles are congruent and the perpendicular bisectors of the base and the top coincide in EF.

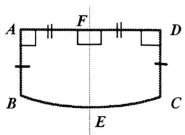

Other arguments can be made using the lunes obtained by extending the four sides of a KQ.

Example of Students' Work on Problem 31

Student answer:
Problem 31 - David Baron

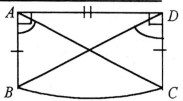

Given: $\angle A = \angle D = \pi/2$; $AB = CD$

Prove: $\angle ABC = \angle DCB$ and perpendicular bisector of BC is perpendicular bisector of AD.

   a. Draw diagonals AC and BD, $AB = DC$; $AD = AD$; $\angle BAD = \angle CDA$, then $\triangle BAD \cong \triangle CDA$ by SAS.

   $AC = BD$ by CPCTE, as well as $\angle BDA = \angle CAD$ by CPCTE.

   And since $\angle BAD = \angle CDA = \pi/2$, $\angle BAD - \angle CAD = \angle CDA - \angle BDA$ but then $\angle BAC = \angle CDB$.

   Hence $\triangle BAC \cong \triangle CDB$ by SAS and $\angle ABC = \angle DCB$ by CPCTE.

   b.

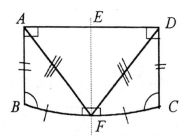

Prove: EF is perpendicular bisector of AD.

   It was proved above that $\angle B = \angle C$, therefore, $\triangle ABF \cong \triangle DCF$ by SAS. Hence $AF = DF$ by CPCTE. $\angle AFE = \angle DFE$ because $\angle BFE = \angle CFE = \pi/2$, so $\angle BFE - \angle BFA = \angle CFE - \angle DFE$, the two smaller angles being equal by CPCTE; $EF = EF$. So $\triangle AFE \cong \triangle DFE$ by SAS. Therefore, $\angle AEF = \angle DEF$ by CPCTE. Hence they both equal to $\pi/2$. Likewise, $AE = DE$ by CPCTE. So FE must be the perpendicular bisector of AD.

PROBLEM 32. *Dissecting Spherical Triangles*

Show that every small spherical triangle is equivalent by dissection to a Khayyam Parallelogram with the same base.

Let us look at dissections of triangles on the plane from which it is possible to build a proof of equivalence by dissection to a parallelogram:

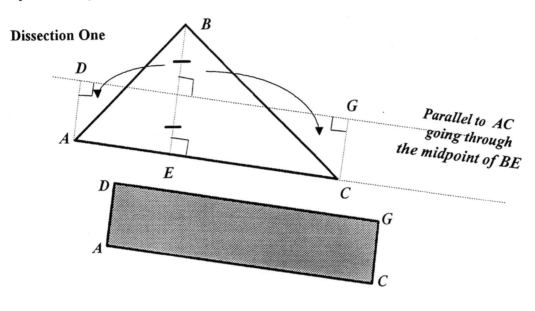

Dissection One

Parallel to *AC*
going through
the midpoint of *BE*

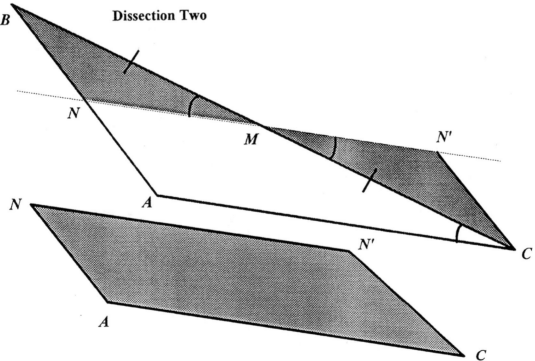

Dissection Two

Dissection Three

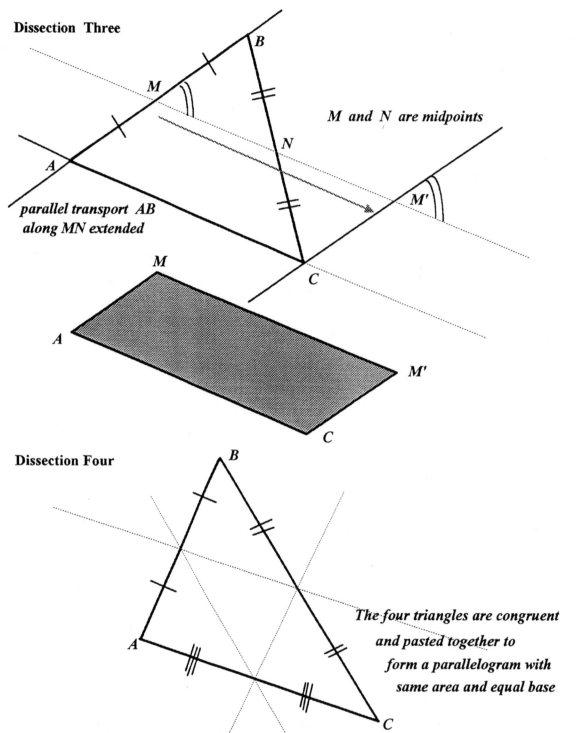

M and N are midpoints

parallel transport AB along MN extended

Dissection Four

The four triangles are congruent and pasted together to form a parallelogram with same area and equal base

Each dissection will require a different proof. Again, we are trying to prove that the result of the dissection is a parallelogram in the plane. Any needed properties of parallel lines or parallelograms should and can be easily proved using what we have already proved in previous problems. Dissection Three can be used to build a proof on the plane as well as on a sphere, however, the two proofs will be different. Let us look at a proof on a sphere using Dissection Three.

Proof:

Using the definition of parallel transport, VAT, and SAA (if we take triangles with sides less than 1/2 great circle), it can be shown that $\triangle MNB \cong \triangle M'NC$. Consequently, $MB \cong M'C$, and $MA \cong M'C$, since $MB \cong MA$. So, $[MACM']$ is a KP.

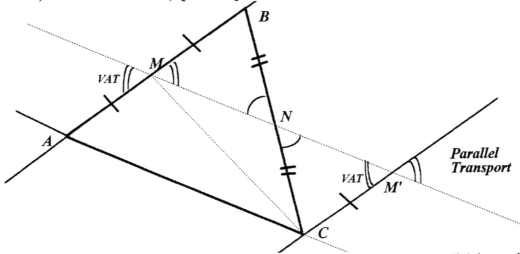

On the plane, we also have to prove that MM' is parallel to AC. (That AM is parallel to CM' follows from the parallel transport and from Problem 16.) Consider the diagonal of the $[MACM']$ that joins M to C. Two triangles are formed, namely, $\triangle MAC$ and $\triangle CM'M$. Given that the diagonal cuts AM and CM', which are parallel transports, we know that it cuts them at congruent angles which implies that $\angle MCM' \cong \angle CMA$. Therefore, $\triangle MAC \cong \triangle M'CM$, and thus $MM' \cong AC$ and $\angle MAC \cong \angle CM'M$. By Problem 16, then, AC is parallel to MM'. Thus, we have proven that $[MACM']$ is a parallelogram.

The three other dissections illustrated above produce parallelograms on the plane, but they do not work on a sphere.

Example of Students' Work on Problem 32

Student answer:

Problem 32 - Christos Ioannou

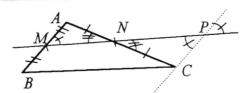

We have $\triangle ABC$. We connect M (midpoint of AB) and N (midpoint of AC) we draw PC - parallel transport of AB along MN going through C. We compare $\triangle AMN$ with $\triangle NPC$:

$\angle AMN = \angle NPC$, $\angle ANM = \angle CNP$, $AN = AC$ by SAA, and since these triangles are small, we have $PC = BM$, and thus we have a *KP*.

PROBLEM *33. Dissecting Khayyam Parallelograms*

Prove that every Khayyam Parallelogram is equivalent by dissection to a Khayyam Quadrilateral with the same base.

Recall the statement of Problem 30:

Show that on the plane every parallelogram is equivalent by dissection to a rectangle with the same base and height

We will prove these two results together, although when one sees them for the first time it is easier to look at the planar case first. The same general approach works on both the plane and a sphere, but one must be more careful on a sphere.

Proof:

There are two cases depending on whether or not a line (geodesic) drawn from one of the two base vertices perpendicular to the top, (extended, if necessary) intersects between A and D.

When the perpendicular does intersect the top we can dissect the parallelogram through that line. If we call the intersection of the perpendicular with the top L, we form $\triangle CLD$, which is a right triangle. If we parallel transport $\triangle CLD$ along AD enough to make the vertex C coincide with the point B, we form the quadrilateral $[L'BCL]$. Since DC and AB are congruent and parallel transports along AD, it is clear that the transported triangle fits and that $[L'BCL]$ is a KQ.

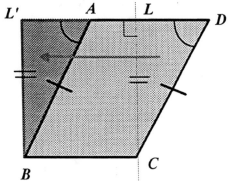

Rather than using a parallel transport argument, one could show $\triangle LDC \cong \triangle L'AB$ by SAA, by raising a perpendicular from B to L' on AD and extending it.

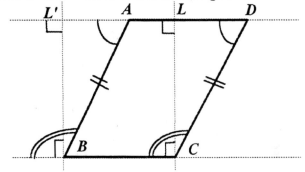

Dissecting Long Skinny Parallelograms

In this problem, students have the tendency to disregard the requirements *same base* or *same base and height*. Thus, they will often disregard the case where the perpendicular at one of the vertices of the base does not intersect the top of the parallelogram or KP. Here is such a case:

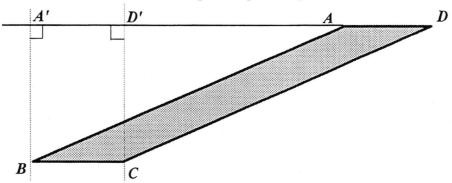

The figure below depicts one dissection on the plane for this type of parallelogram. After tracing the perpendicular, call L' the intersection of the line with the side of the parallelogram. Trace the line parallel to the base passing through L'. This line is going to intersect the other side of the parallelogram at a point, P'. Raise a perpendicular to this line which will intersect the other side of the parallelogram, at the point L''. Do as many cuts as are necessary until one of the lines cuts the top of the parallelogram. This last piece of the original parallelogram can be dissected as illustrated below.

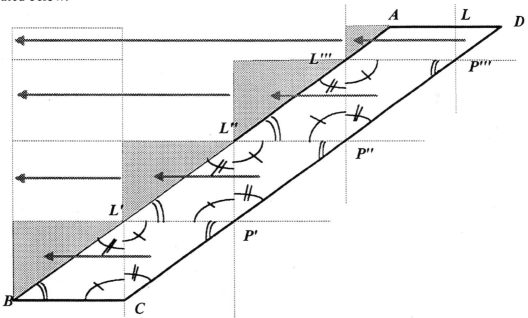

For every two consecutive cuts (one perpendicular to the base and another parallel to the base) we cut a piece that is a triangle. The triangles are right triangles that are congruent to each other and even in number. If they are collected in groups of two and the hypotenuses made to coincide the "right way," we obtain a rectangle for each pair of triangles. These rectangles can then be stacked in a pile that will form a rectangle with the same base and height as the original parallelogram. Note that since the triangles all have corresponding congruent angles and any two consecutive triangles always share one side, they can be proven to be congruent "in a chain" using ASA.

Why is the number of steps finite? If we assume the Archimedean property of the real numbers, then if the height of the original parallelogram is h, the length of $L'C \cong L''P' \cong L'''P''$ is a, and there exists a natural number n such that $na > h$. The process can be completed, then, with $n-1$ steps.

This dissection will not work, as described above, on a sphere. A careful modification of it, however, will work on a sphere. In the figure below, consider the geodesic that contains the top of a KP and call it g. Consider the two geodesics that go through B and C and are perpendicular to g. We have formed a four-sided polygon, namely, $[A'BCD]$. Along the geodesic g parallel transport both polygons, the KP and $[A'BCD]$, until copies of the KP cover $[A'BCD]$ and vice versa. In the figure below we have 5 pieces, one of them common to the two polygons. For each of the other four pieces in the KP there exists a corresponding congruent piece in the KQ.

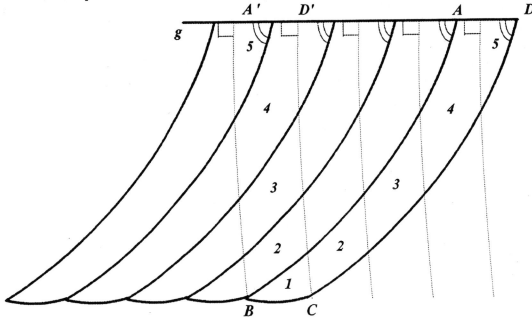

There are at least two other proofs that work for all parallelograms whether planar or spherical and whether long and skinny or not.

Diagonal proof:

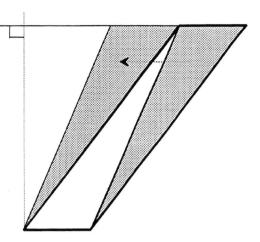

Repeat until the top
contains the dotted
perpendicular and then
cut along the dotted
line and form a
rectangle.

Subtraction proof:

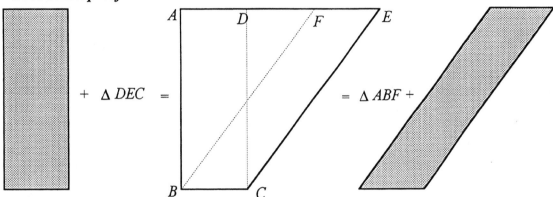

$+ \ \Delta DEC \ =$ $= \ \Delta ABF \ +$

Since $\Delta DEC \cong \Delta ABF$, the parallelogram and the rectangle are equivalent by substraction. Notice that this proof holds even if the base is infinitesimal in length.

Example of Students' Work on Problem 33

Professor and T.A.   *Student answer:*
comments:

Problem 33 - Emma Lister

Take a KQ, *ABCD*:
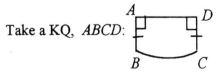

Draw dotted lines so that they dissect the KQ like so:

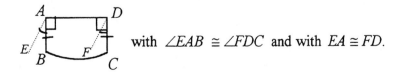   with $\angle EAB \cong \angle FDC$ and with $EA \cong FD$.

Therefore, by property of right angles:

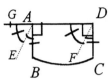 we have that $\angle GAE \cong \angle ADF$.

Then we can connect point E to point B and point F to point C to form triangles $\triangle AEB$ and $\triangle DFC$:

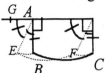 And so, the picture shows that SAS proves
$$\triangle AEB \cong \triangle DFC.$$

OK, but I don't see how the area can be dissected.

Therefore by corresponding parts of congruent triangles, $EB \cong FC$. The distance between point B and point C is equal to the distance between point E and point F. In a sense that on the sphere, we can move BC over so that it can be the new base EF of the KP $AEFD$.

Why is it congruent to BC?

Response to comments:

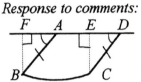

Take KP, $ABCD$. Draw a line CE.
Parallel transport line CE along AD to BF.
By SAA, $\triangle BFA \cong \triangle CED$. (for small triangles on the sphere).
And then KQ:$BFEC$ is a dissection of KP:$ABCD$ with same base.

Can you do the reverse?

PROBLEM 34. *Spherical Polygons Dissect into Biangles*

[This problem is not used elsewhere in the book and thus may be omitted.]

In the next chapter, students will show that every polygon on the plane is equivalent by dissection to a square. This does not apply to the sphere because there are no squares on the sphere. However, we have already shown in Problem 10 that two polygons on the same sphere have the same area if they have the same holonomy. Thus, every polygon on the sphere must have the same area as the biangle with the same holonomy. Now we can show that not only do they have the same area, but they are equivalent by dissection.

Show that every small simple polygon on a sphere is equivalent by dissection to a biangle with the same holonomy. I.e. the angle of the biangle is equal to $(\frac{1}{2})(2\pi - $ Sum of the exterior angles of the polygon).

Consequently, two small simple polygons on the sphere with the same area are equivalent by dissection.

Proof:
The proof of this result can be completed by proving the following Lemmas:

Lemma 1.

Every small simple polygon can be dissected into a finite number of small triangles, such that the holonomy of the polygon is the sum of the holonomies of the triangles.

This follows immediately from Problems 10 and 11.

Lemma 2.

Each small triangle is equivalent by dissection to a KQ with the same base and the same holonomy.

Check over any proofs of Problems 32 and 33.

Lemma 3.

Two KQ's with the same base and the same holonomy (or base angles) are congruent.

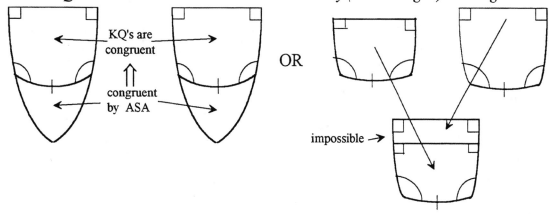

Lemma 4.

If two triangles have the same base and the same holonomy, then they are equivalent by dissection.

Combine the previous two lemmas.

Lemma 5.

Any triangle is equivalent by dissection to a biangle with H (Δ) = H *(biangle)* = twice the angle of the biangle.

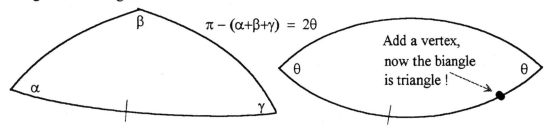

Therefore, any small simple polygon can be dissected into triangles and each dissected in a biangle. The biangles can then be put together to form one combined biangle with the same holonomy as the polygon. Thus two polygons with the same holonomy are equivalent by dissection to the same biangle and thus equivalent by dissection to each other.

Chapter 13

Square Roots, Pythagoras, and Similar Triangles

Problems 35 through 37 allow students to make connections between Geometry and Algebra through the concept of area. In Problem 35 students are asked to prove a theorem that may well be more than 3700 years old!

PROBLEM 35. A Rectangle Dissects into a Square

Show that, on the plane, every rectangle is equivalent by dissection to a square.

As discussed in the students' manual, this problem appears in the ancient Sanskrit book *Sulbasutram* ("*Rules of the Cord*") by Baudhayana [A: Baudhayana], which was written as a handbook for people who were building altars and temples. Most of the book gives detailed instructions on design and construction, but the first chapter contains explicit directions about how to perform various geometric constructions, including the construction which converts a rectangle into a square. It also contains a clear statement of the Pythagorean Theorem. A. Seidenberg in an article entitled *The Ritual Origin of Geometry*, [G: Seidenberg] gives a detailed discussion of the significance of the Sulbasutram. He argues that it was written before 600 BC (Pythagoras lived about 500 BC, and Euclid about 300 BC). He gives evidence to support the claim that it contains codification of knowledge going "far back of 1700 BC" and that this knowledge was the common source of Indian, Egyptian, Babylonian and Greek mathematics.

At first, students should try this problem on their own. Later, you may want to suggest some strategies for tackling it. Students may find a procedure that involves infinitely many steps — use this as an opportunity to discuss completeness and the real numbers. Or some students may find a procedure that, in general, only works for certain rectangles (for example, any rectangle with one side an integral multiple of the other). They discuss the connections between rational and irrational or the Greek notion of incommensurate. You should point out that it is possible to do the dissection in a few steps, and then suggest techniques for constructing what will be the side of the square. Let us look at some of these hints:

First Construction: Extend the base of the rectangle a, by a length b, the other side of the rectangle. Then, draw a semicircle with diameter equal to the sum of the lengths of the base and side, $a + b$. Extend the side of the rectangle which is not tangent to the semicircle, until it intersects the semicircle. This line goes from the base of the rectangle to the semicircle, and is the side of the square we need to find. Now, complete the square and prove that it can be dissected to form the original rectangle.

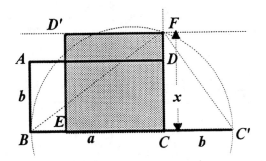

Proof:

Consider $\triangle BFC'$, and notice that $\angle BFC' \cong 90°$ since it is an angle subtended in a semicircle (this result can be proved using Problems 8 and 19). All of the congruent angles in the construction are represented in the figure below. All of these the congruencies follow from the fact that $\angle BFC' \cong 90°$ and from what has been proven in previous chapters about parallel lines and transversals.

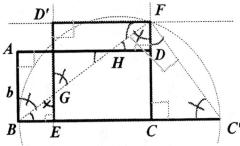

We can find congruent triangles in the following way: By construction, $D'F \cong CF$, which means that $\triangle GD'F \cong \triangle C'CF$ by ASA. But, $\triangle C'CF \cong \triangle BAH$ since $C'C \cong BA$. Therefore, $\triangle BAH \cong \triangle GD'F$. Since $EG \cong DF$ then, $\triangle BEG \cong \triangle HDF$ also by ASA. In the dissection we have constructed, $\triangle BAH$ slides along BF until it coincides with $\triangle GD'F$, and $\triangle BEG$ slides along BF until it coincides with $\triangle HDF$.

Second Construction: Consider the rectangle $[ABCD]$ with sides a and b where $a > b$. Trace the circle that has a as its diameter and the midpoint of BC as its center. Trace a second circle of radius b and center C. Let D' be the intersection of the second circle with BC. Raise a perpendicular from D' to the circle. Let D'' be the intersection of that perpendicular with the circle. $D''C$ is the side of the square we would like to find.

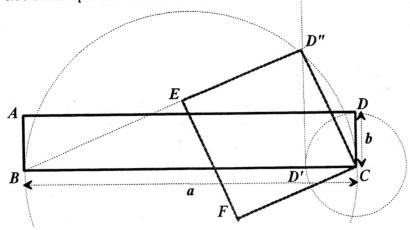

Proof:

Draw a segment perpendicular to BC going through F. (Refer to the diagram below.) Two triangles, $\triangle FF'C$ and $\triangle FF'G$, are formed. Triangles $\triangle FF'C$ and $\triangle CD'D''$ are congruent since their corresponding angles are congruent, and $D''C \cong FC$, by construction. Since $DC \cong BA$, and $DC \cong D'C$, we can conclude that $\triangle CD'D'' \cong \triangle BAH$. Consequently, $\triangle FF'C \cong \triangle BAH$, and these areas, labeled *II* on the figure below, can slide onto each other. Since $FF' \cong CD$ we can show that $\triangle FF'G \cong \triangle CDL$ by ASA, and so the areas labeled *I* can slide onto each other. This latter congruency implies that $D'L \cong EG$ thus $\triangle HD''L \cong \triangle BEG$. But then we can slide the areas labeled *III* onto each other, and the triangle $\triangle HEJ$ that is outside the rectangle is congruent to the uncovered triangular part of the original rectangle.

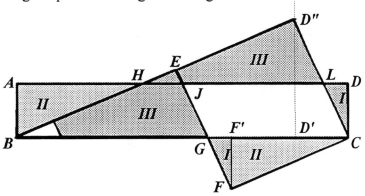

Third Construction: This is the construction that appears in Baudhayana's ***Sulbasutram***. This procedure for dissecting a rectangle into a square uses a description of completing a square, as well as what we now refer to as the Pythagorean Theorem (Problem 36).

Consider the rectangle with sides a and b where $b > a$. Divide the rectangle into a square, [$ABCD$], of side a and a rectangle, [$DCEF$], of sides $b-a$ and a. Divide CE in half, and raise a perpendicular, dividing the rectangle [$DCEF$] into two other congruent rectangles with sides $(b-a)/2$ and a. Move one of these rectangles under the square [$ABCD$] such that the sides of length a coincide. Now, we can ***complete the square*** [$AKHJ$], with side $(a+b)/2$, by adding the square [$CGHI$], with side $(b-a)/2$.

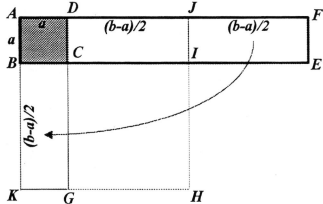

The following relationship results:

$$\text{square on } (a+b)/2 =_d \text{ original rectangle } + \text{ square on } (b-a)/2.$$

Using Problem 36, we can construct the side of the square we are trying to find:

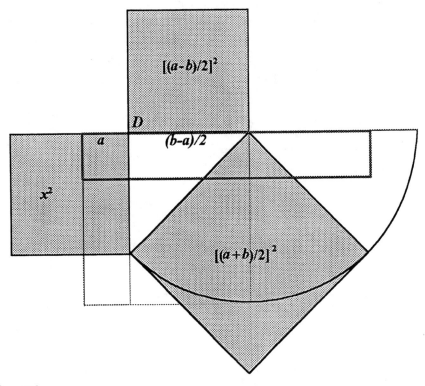

By Problem 36:

$$[(a+b)/2]^2 =_d x^2 + [(b-a)/2]^2$$

and above we showed that:

$$[(b+a)/2]^2 =_d \text{ original rectangle} + [(b-a)/2]^2$$

which, by definition, gives us:

$$\text{square on } x =_s \text{ original rectangle.}$$

Example of Students' Work on Problem 35

Professor and T.A. comments:

Student answer:

Problem 35 - Semyon Kruglyak

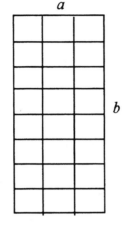

a

b

Given a parallelogram with sides $a, b,$ we can dissect it into ab squares of side 1. This is done by making b-1 cuts of equal width along b and a-1 cuts of equal width along a . Since we are dividing the sides into $1/a, 1/b$ segment, each segment has length $(1/a) \cdot a, (1/b) \cdot b = 1$. Since the units are arbitrary, we can always pick them such that a, b are integers. For example, if $a = 2$ cm, $b = 3.5$ cm, we simply use units called 1/2 cm and $a = 4$ (1/2 cm), $b = 7$ (1/2 cm).

How can you be sure that you can find a common measure (unit) for a and b?

Using # 29, we combine 2 of the small squares into a bigger square, then take the bigger square and combine it with another square. This process may be continued until all ab squares are combined into one square.

Response to comments:

My solution was to divide the rectangle into ab 1×1 squares where 1 is a common unit for a, b . The only time this fails is if one of the sides is irrational. For example if $a = 1$ and $b = 3.14$, we can say $a = 100$, $b = 314$ of a different unit. Unfortunately, if $a = 1$ and $b = \pi$ this will not work.

However, all lengths are measured to a certain accuracy and in practice a line of length π cannot be drawn. Any approximation of π , $\sqrt{2}$, etc. is rational and my method will work. I still tried to find another method to convert the rectangle into a square. I came up with the following:

Nice

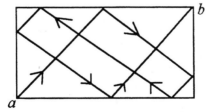

b

a

Start at a and continue at 45° angles until b is reached. This will divide the rectangle into squares and an even number of isosceles right triangles which can easily be put together to make a square.

Good.

This method looked good for a while but I think that it is based on the idea that the side lengths have a factor in common. If a side has irrational length, I don't think that b will ever be reached. Finally I tried the construction given in class, but this was as far as I got:

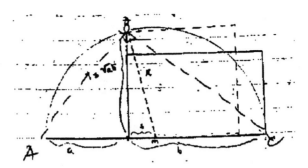

$m = (a+b)/2 =$ radius (r); $s = b-(a+b)/2 = (b-a)/2$; $t =$ length of side of square.

Response to comments:
Using the following construction;

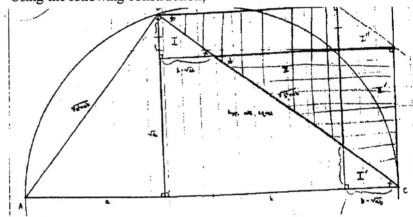

it is only necessary to show that $\Delta I \cong \Delta I'$ and $\Delta II \cong \Delta II'$. By ASA, $I \cong I'$. Both are right triangles, the base of each is $b-\sqrt{ab}$, the angles to the right of the base are congruent because a line cuts parallel lines at congruent angles. If it is not clear that the base of the top triangle = $b-\sqrt{ab}$, it can be easily be shown that $I \cong I''$ and $I' \cong I''$ and then $I \cong I'$. $II \cong II'$ by SAA. The hypotenuse of $II =$ that of II', because one consists of BC-hypotenuse (I) and the other of BC-hypotenuse (I'). Since $I \cong I'$, hypotenuse (I) = hypotenuse of (I'). The angle equivalence comes from BC intersecting parallel lines, triangles both have angle 90°. Since we can cut part of the rectangle off and rearrange the pieces to form a square, the two figures are dissection equivalent.

PROBLEM 36. *Equivalence of Squares*

Prove the following: *On the plane, the union of two squares is equivalent by dissection to another square.*

Proof:

Consider two squares with side lengths a and b. Position one square above the other such that one vertex and two edges line up, as illustrated below. Extend that geometric figure to a square of side length $a+b$. Let I be the intersection of the extension of AE with the square of side $a+b$. GI is the side of the square we are looking for. Trace the congruent (by SAS) triangles, $\triangle GHI$, $\triangle GFD$, $\triangle IAB$, $\triangle DCB$, in figure below. The quadrilateral, $[IBDG]$, formed by these triangles has sides equal to the hypotenuses of these triangles. $[IBDG]$ is a square because each of its angles is the sum or the supplement of the two non-right angles in the right triangle $\triangle GHI$. The dissection of $[ABCE]$ plus $[GHEF]$ into $[IBDG]$ then becomes evident.

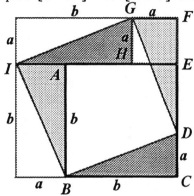

Looking at this figure we also see the geometric representation of the expressions:

$$(a+b)^2 = a^2 + b^2 + 2ab$$

and,
$$(a+b)^2 = c^2 + 2ab,$$

which lead to
$$c^2 =_s a^2 + b^2.$$

Another proof:

Consider the triangle $\triangle ABC$ and construct on its sides the squares $[GBCD]$ and $[EFAB]$. Reflect the square $[AIJC]$ through AC to $[AI'J'C]$. Trace a line perpendicular to GB from I' to H. Then, since $AC \cong I'A$, $\triangle I'HA \cong \triangle ABC$ by ASA and thus $I'H \cong AB \cong EB$. Slide $[EFAB]$ along AH until EB coincides with $I'H$. From this we get congruencies between triangles that give us the dissection of the squares into another square. See the illustration below.

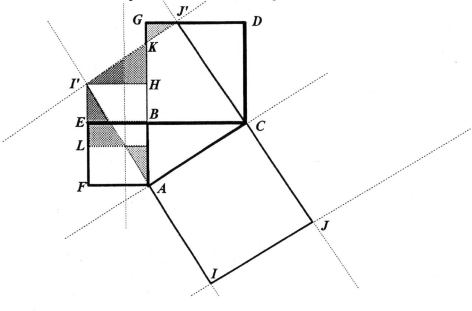

Still another proof:

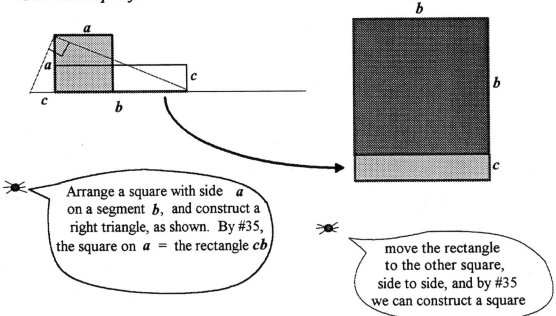

Arrange a square with side **a** on a segment **b**, and construct a right triangle, as shown. By #35, the square on **a** = the rectangle **cb**

move the rectangle to the other square, side to side, and by #35 we can construct a square

Example of Students' Work on Problem 36

Professor and T.A. comments:

Student answer:

<u>**Problem 36 — Jeremy W. Schulman**</u>
<u>1st Version:</u>

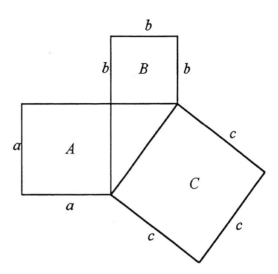

By Pythagorean theorem, $a^2 + b^2 = c^2$, specific form of law of cosines:

$$c^2 = a^2 + b^2 - 2ab \cos 90°$$
$$c^2 = a^2 + b^2 - 0$$
$$c^2 = a^2 + b^2$$

Isn't the Pythago-rean Theorem what you have to prove?

The area of square C is c^2.
The area of square B is b^2.
The area of square A is a^2.

Good idea.

A^2
B^2

O.K., but add a few more specifics, e.g. which rectangles?

By 35, we can dissect square B to form a rectangle. We can dissect square A to form a rectangle. These two rectangles, again by 35, join to form square C.

Nice start. Can you now, using congruency, prove that the areas are the same?

2nd Version:
We can see how the dissection of squares occurs:

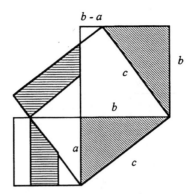

Response to comments:

<u>1st Version:</u>

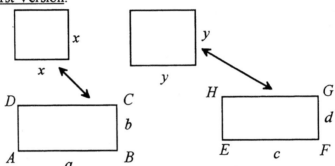

Do Problem 35 in reverse. Create rectangles $ABCD$ and $EFGH$ such that $a = c$.

i.e.

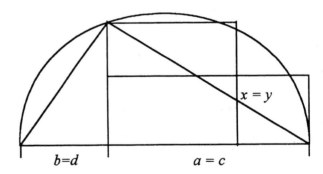

Basically, we must stipulate that $a + b = c + d$. If we keep $a = c$, then b and d will differ.

Therefore, since $a = c$, we can do this:

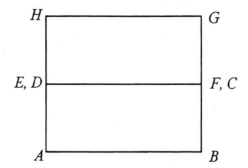

Rectangle *ABGH* can be dissected into a square by Problem 35. Therefore, any two squares are equivalent by dissection to a bigger square.

<u>2nd Version</u>:

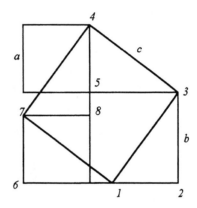

By ASS, $\Delta 123 \cong \Delta 453$ and also by ASS, $\Delta 761 \cong \Delta 784$. Therefore, any two squares $=_d$ bigger square.

PROBLEM 37. Similar Triangles

Students can re-investigate Problem 35, using a dissection theory proof to show that:

If two triangles have corresponding angles congruent, then the corresponding sides of the triangles are in the same proportion to one another.

Proof:

Let θ be one of the angles of the triangles, and place the two θ's in VAT position (see the figure below). Parallel transport $\triangle ABC$, first along BD and then along DE. Then $A'C''$ is a straight line, and if we trace a line through B parallel to DE, we obtain two more triangles that are congruent to $\triangle ABC$. Note, then, that the two remaining pieces of the parallelograms $[AA'DB]$ and $[BB''C''C]$ have congruent bases, as well as the same height, and thus, by Problem 30, they are equivalent by dissection. In formulas

$$[AA'DB] =_d [BB''C''C].$$

By the discussion after Problem 36 in the student's text, we conclude:

$$\text{Area}(\,[AA'DB]\,) = \text{Area}(\,[BB''C''C]\,).$$

Now, if $\theta = \pi/2$, then the above equal areas are the products $ab = cd$, and cross-dividing we get the desired $a/d = c/b$. Then we can define the usual trigonometric functions for right triangles. For arbitrary θ the above equal areas are $ab \sin\theta = cd \sin\theta$, again cross-dividing, we get:

$$a/d = c/b\,.$$

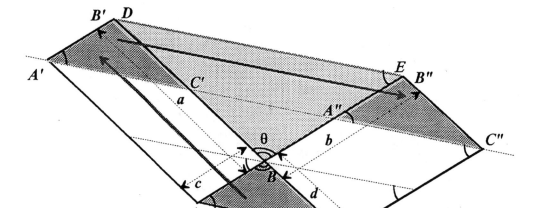

Chapter 14

Geometric Solutions of Quadratic and Cubic Equations

I urge you to not look at this chapter only for its historical interest but rather look for the meaning it has in our present-day understanding of mathematics.

PROBLEM 38. *Quadratic Equations*

I have not used this problem, as stated in my course, and thus I do not have any experience as to how students will respond. However, I do know that most of my students have expressed an interest in this topic and have explored it on their own.

I suggest that you read what is said in [G: Joseph] and [G: Eves] about solutions of quadratic equations. It is especially important to read Joseph's book because most historical account of mathematics slight or ignore the contributions of non-Western cultures. I would also suggest bringing into the classroom copies of Omar Khayyam's *The Rubaiyat* and translations of the two books entitled *Al-jabr w'al mugabalah,* [A: al'Khowarizmi] and [A: Khayyam, 1931]. In addition you will probably find it helpful to have available [A: Khayyam, 1963] and [A: Cardano].

a. *Show that these are the only types. Why is* $x^2 + bx + c = 0$ *not included? Explain why* b *must be a length but* c *an area.*

The trick here is that they thought of x, b, c as geometric quantities and not as real numbers in the modern day sense.

b. *Speculate about why mathematicians avoided negative numbers?*

In other courses where I decide to spend more time on the underlying meanings of numbers and arithmetic I assign the following sequence of problems. These problems do succeed in getting the students to examine carefully what the meanings are for them. You may find it helpful to use some of these problems in your discussions on Problem 38.

A. *Why does Analyzer** [many other graphing software would work also] *graph* $x^{50} + 1 - x^{50}$ *as it does?*

B. *Each of the following problems involve the same long ribbon:*

1. *Divide the ribbon equally among 4 people. How much ribbon does each person get?*

2. *How many packages can be tied with this ribbon if each package needs 4 feet of ribbon?*

3. *Someone makes a scale model of the ribbon at a ratio of 1:4. How long is the model?*

If L denotes the length of the ribbon, then the numerical answer to each problem is L/4.

i. *Why should these very different problems have the same numerical answer?* Notice that each problem implies a very different physical process.

ii. *Give a definition (or a method of construction) of the product of two positive integers.*

iii. *Assume L = 36 feet. Relate (as concretely as you can) the answers and processes in each problem to your defined meaning of* $9 \times 4 = 36$ *or* $4 \times 9 = 36$.

C. *Assume (from* **A.***) that we have defined, for a, b positive integers and C, D quantities:*

Scalar multiplication, $a \cdot C$.

Arithmetic multiplication (or multiplication of scalars), $a \times b$.

Addition of quantities and positive integers, $C + D$ or $a + b$.

Find as many properties as you can that these operations satisfy (e.g. $a \cdot (b \cdot C) = (a \times b) \cdot C$ and $(a+b) \cdot C = a \cdot C + b \cdot C$). *Show how you know that each is true.*

****

Assume that $a \cdot C$ and $a \times b$ has been defined for all lengths C and rational numbers (fractions) a and b. Previously in the course we saw that the positive reals, R^+, (scalars, "pure" numbers) defined in three different ways:

i. $R^+ = R^+_r = $ ***ratios of geometric lengths***, where $A:B = C:D$ iff (a rectangle of sides A and D has the same area as a rectangle with sides C and B) **or** (a right triangle with legs A and B is similar to a right triangle with legs C and D).

ii. $R^+ = R^+_s = $ ***limits of sequences of positive rationals***, where $\lim\{a_i\} = \lim\{b_i\}$ iff $\{a_i - b_i\} \approx 0$ (i.e., for each positive integers N, $|a_i - b_i| < 1/N$, for all sufficiently large i.).

iii. $R^+ = R^+_d = $ (ordinary) ***positive decimal expansions*** where $.999\cdots = 1.000\cdots$.

****

D. *Show how to define* $a \cdot C$ *and* $a \times b$ *for a,b in* R^+_r *and C a length.*

Are your definitions "well-defined"?

[I.e. if $a = c$ why is $a \cdot C = c \cdot C$ and $a \times b = c \times b$?]

E. *Define* $a \cdot C$ *and* $a \times b$ *for C a length and*

i. *a,b in* R^+_s.

ii. *a,b in* R^+_d.

Are your definitions "well-defined"?

[I.e. if $a = c$ why is $a \cdot C = c \cdot C$ and $a \times b = c \times b$?]

F.   **i.** *How do you define multiplication (scalar and arithmetic) of negative numbers as naturally as possible so that* 2×-3 *and* -2×3 *have different meanings?*

ii. *What about multiplication by zero?*

G. *Let* L *be a geometric line with an origin identified.*

i. *Show that* L *is in a natural way a vector space* (over R). [Use standard properties from Problems 2-6.]

ii. *What is a basis for* L *? Why?*

c. *Find geometrically the algebraic equations which express all the positive roots of each of the six types. Fill in the details in the following sketch of Khayyam's methods for Types 3-6.*

Look in [**A:** al'Khowarizmi], [**A:** Khayyam, 1931], and [**A:** Khayyam, 1963] for further discussions of these solutions if necessary.

Do the above solutions find the negative roots? Well, first, the answer is clearly, "No," if you mean: "Did al'Khowarizmi and Khayyam mention negative roots?" But let us not be too hasty. Suppose $-r$ (r, positive) is the negative root of $x^2 + bx = c$. Then $(-r)^2 + b(-r) = c$ or $r^2 = br + c$. Thus r is a positive root of $x^2 = bx + c$! The absolute value of the negative root of $x^2 + bx = c$ is the positive root of $x^2 = bx + c$ and vice versa. Also, the absolute values of the negative roots of $x^2 + bx + c = 0$ are the positive roots of $x^2 + c = bx$. So, in this sense, *Yes, the above geometric solutions do find all the real roots of all quadratic equations.* Thus, it is misleading to state, as most historical accounts do, that the geometric methods failed to find the negative roots. The users of these methods did not find negative roots because they did not conceive of them. However, the methods can be directly used to find all the positive and negative roots of all quadratics.

d. *Use Khayyam's methods to find all roots of the following equations:* $x^2 + 2x = 2$, $x^2 = 2x + 2$, $x^2 + 3x + 1 = 0$.

PROBLEM 39. Conic Sections and Cube Roots

It is important that you make sure that the students use good (accurate) graph paper and make some careful constructions of hyperbolas, parabolas, and cube roots. Practice is necessary.

a. *Use the above geometric methods with a fine graph paper to find the cube root of* 10.

b. *Use the above method with graph paper to construct the graph of the hyperbola with parameter 5. What is an algebraic equation that represents this hyperbola?*

PROBLEM 40. Roots of Cubic Equations

a. *Show that in order to find all the roots of all cubic equation we need only have a method that finds the roots of Types 1, 2, and 3.*

b. *Verify that Khayyam's method described above works for Types 2 and 3. Can you see from your verification why the extraneous root given by B appears?*

c. *Use Khayyam's method to find all solutions to the cubic*

$$x^3 = 15x + 4.$$

Use fine graph paper and try for three-place accuracy.

I find that students have very little trouble with this problem **unless** they do not actually make the constructions. Some students need extra encouragement because they have been conditioned to avoid constructions.

PROBLEM *41. Algebraic Solution of Cubics*

a. *Solve the cubic* $x^3 = 15x + 4$ *using Cardano's Formula and your knowledge of complex numbers.*

The relations $t + u = -b$ and $t^{1/3} u^{1/3} = -(a/3)$ can be used in a straightforward way to find which of the nine possible values in $x = \{2 + \sqrt{-121}\,\}^{1/3} + \{2 - \sqrt{-121}\,\}^{1/3}$ are actually solutions. This is because the product of the cube roots must be the real number $-(a/3)$ their arguments must add up to a multiple of π.

b. *Solve* $x^3 = 15x + 4$ *by dividing through by* $x - 4$ *and then solving the resulting quadratic.*

If your students know other methods for finding solutions then assign them also. For example, the roots may be found by iteration techniques, such as Newton's Method.

c. *Compare you answers and methods of solution from Problems* 40**c**, 41**a**, *and* 41**b**.

This is an opportunity for the students to reflect back on what they have done. You will probably find a large divergence of feelings and observations here — this will lead to a good class discussion.

Chapter 15

Projections of a Sphere onto a Plane

In Problems 42, 43 and 44 students have a chance to explore tools that allow them to relate the geometry of a sphere with the geometry of the plane. These tools map a piece of a sphere's surface onto a plane.

PROBLEM *42. Gnomic Projection*

Show that gnomic projection takes the portions of great circles in the lower hemisphere onto straight lines in the plane.

In the problems dealing with projections, students must think extrinsically. One hint you may want to give students is that defining a great circle extrinsically is one step toward thinking about a spherical surface extrinsically. You may also want to tell the students that they can think of gnomic projection physically. A physical way of looking at gnomic projections is to imagine a light source at the center of a transparent sphere. The rays of light will connect, in a straight line, the center of the sphere, any point P on the lower hemisphere, and a unique point on the plane Π tangent to the sphere at the south pole. The latter point is called the gnomic projection of P.

Proof:

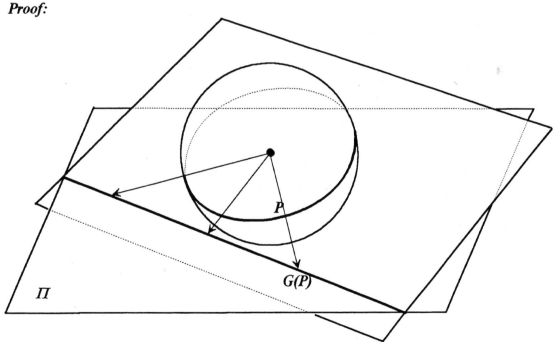

A great circle is determined by a plane through the center of the sphere. The intersection of this plane with the plane Π is a straight line which is the gnomic projection of this great circle.

Example of Students' Work on Problem 42

Student answer:

Problem 42 - Michael Kraizman

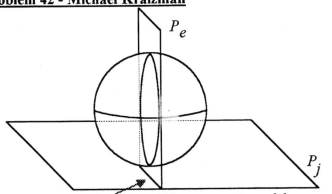

Line formed by intersection of the two planes

Good

We can see that a great circle describes a plane that cuts through (bisects) the sphere along every point on the great circle. Taking the intersection of this pane with the projection plane, we get a straight line. Since we can do this for any lower half of a great circle, we see that all great circles are projected onto straight lines.

Note that a Gnomic projection is a chart only for half of the sphere.

PROBLEM *43. Cylindrical Projection*

Show that cylindrical projection preserves area.

In cylindrical projection, all points on a sphere are projected onto a cylinder with the same radius as the sphere and a height equal to its diameter. Cylindrical projection can be used to compute area on the sphere because it preserves area.

Geometric Proof:

Consider a very small element of area on the sphere, bounded by two latitudes and two longitudes. Now project this area onto the cylinder. The appropriate 2-dimensional cross-sections in the drawing below show that the vertical length a is projected onto a segment with length $a \cos(\theta)$ and the horizontal length b is projected onto a segment of length $b/\cos(\theta)$. Thus the image of the small element of area has area $[a \cos(\theta)][b/\cos(\theta)] = ab$.

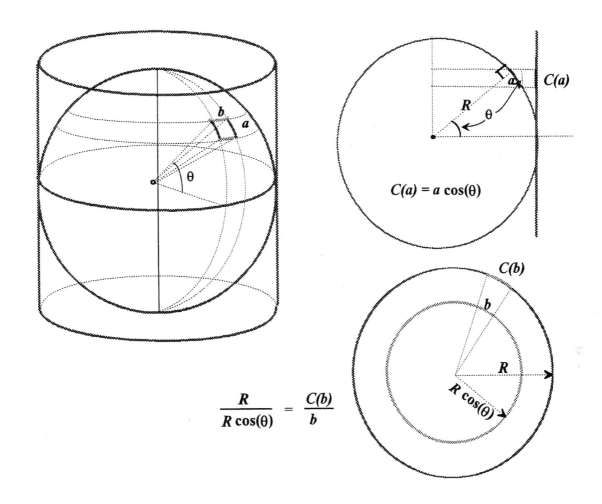

$$\frac{R}{R\cos(\theta)} = \frac{C(b)}{b}$$

Analytic Proof:

Find a function f from a rectangle in the (z, θ)-plane onto the sphere and a function g from the same rectangle onto the cylinder so that $C(f(z, \theta)) = g(z, \theta)$.

$$f(z, \theta) = (\sqrt{R^2 - z^2}\,\cos\theta, \sqrt{R^2 - z^2}\,\sin\theta, z)$$

$$g(z, \theta) = (R\cos\theta, R\sin\theta, z)$$

Let us consider an element of area $dz\, d\theta$. The element of length dz is transformed by f to $f_z dz$, where f_z is the partial derivative with respect to z. For two vectors A, B, the magnitude of the cross product $|A \times B|$ is the area of the parallelogram determined by A and B. Therefore,

$$dz\, d\theta \text{ is transformed by } f \text{ to } |f_z \times f_\theta|\, dz\, d\theta$$

$$\text{and by } g \text{ to } |g_z \times g_\theta|\, dz\, d\theta.$$

But, we can directly calculate that

$$|f_z \times f_\theta| = R = |g_z \times g_\theta|.$$

Thus, C preserves area.

Example of Students' Work on Problem 43

Professor and T.A. comments:

Student answer:

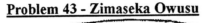

Problem 43 - Zimaseka Owusu

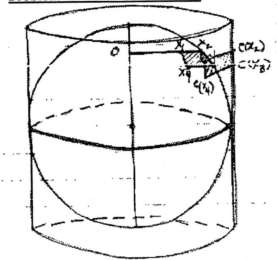

If one takes an infinitesimal area on the sphere, A, as in figure above one gets the cylindrical projected area A' (blown up in figure below):

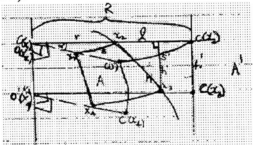

The section of the great circle between x_2 and x_3 can be approximated to an extrinsic line h and by Pythagoras theorem $h > h'$.

Nice.

Prove that in fact they compensate.

The horizontal dimension, on the other hand, $s' > s$ because $s = r\alpha$ and $s' = R\alpha$ and $r < R$ except at equator. So when the infinitesimal area A is cylindrically projected, the horizontal dimension becomes longer while the vertical dimension becomes shorter. These compensate for each other, and to see this, the sphere could be rotated 90° about an axis through the center of the sphere and the center of the area A such that the original horizontal dimension becomes vertical and vice versa and the same area can be projected with the result that the horizontal dimension becomes longer though it had been shorter as vertical and vice versa, though the area is the same.

Response to comments:

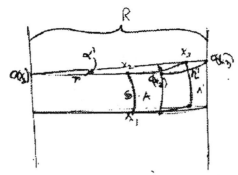

From figure in the previous answer we have that $s=r\alpha$ and $s'=R\alpha$ which implies that $(s/s')=(r/R)$. Rotating the sphere such that h becomes the horizontal component (see figure immediately above) we get $h=r\alpha$ and $h''=R\alpha$, that is $(h/h'')=(r/R)$. Therefore, the horizontal component of the projected area becomes longer, and the ratio of the horizontal component h to the projected horizontal $(h/h'')=(r/R)=(s/s')$ before the rotation. Hence they compensate.

PROBLEM *44. Stereographic Projection*

Show that stereographic projection preserves size of angles.

Thinking physically again, in this case we should imagine the source of light to be at the north pole, N, instead of at the center. Points in a neighborhood of N are mapped "far away" on the plane tangent to the sphere at the south pole, N^*, and points in a small neighborhood of N^* are mapped right under the sphere. This projection does not project all great circles onto straight lines, however, it does project all great circles that go through N and N^* onto straight lines. In Problem 44, students are asked to prove that all stereographic projections are ***conformal maps***, that is, they preserve angles.

Proof:

Let P be the vertex of an angle α on the sphere. Consider the (extrinsic) vectors, v and w, which are tangent to the sides of the angle at P. These vector are contained in the plane T which is tangent to the sphere at P. (See figure below.) The North Pole (N) and the vectors determine two planes, (v,N) and (w,N). The image of α is determined by the intersection of these planes with the image plane Π which is tangent to the sphere at the South Pole.

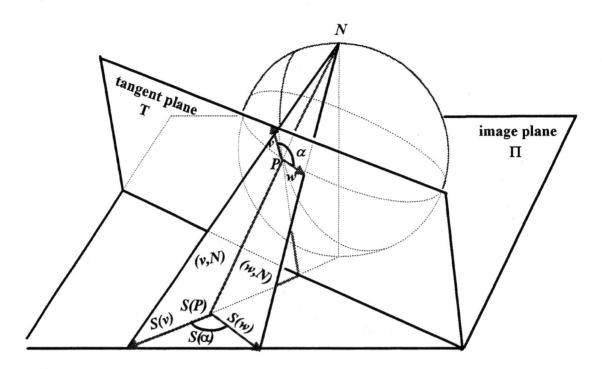

In the following 2-dimensional cross-sectional drawing, we can see that the planes T and Π intersect the line *NP* at the same angle. Thus, the plane which bisects the dihedral angle between T and Π will reflect the intersections of (v,N) and (w,N) with T onto the intersections of (v,N) and (w,N) with Π. Thus, we see from the drawings that α is congruent to $S(\alpha)$.

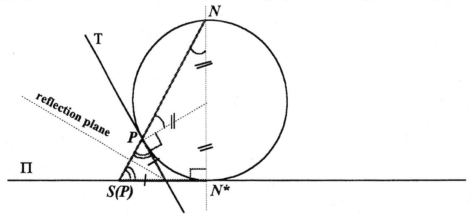

Let us look now at an analytic proof of Problem 44:

Analytic Proof:

First, we must introduce coordinates on the image plane and on the sphere. There are many ways to do this. Using polar coordinates on the plane and spherical coordinates on the sphere, we get a fairly simple formula for the inverse of *S*. Checking that angles are preserved is more complicated, however. The approach that involves Calculus in the most direct way is to use rectangular coordinates in 3-space with the origin at the south pole and with coordinates $(0,0,2)$ at the north pole.

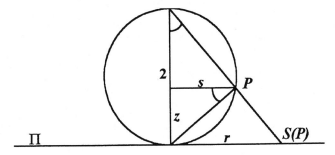

Set $r = \sqrt{x^2 + y^2}$. Using the similar triangles in the drawing above, we see that $\frac{s}{2-z} = \frac{r}{2} = \frac{z}{s}$. Eliminating s, we obtain $z = \frac{2r^2}{4+r^2}$. Thus, the inverse of S is the function:

$$f(x,y) = \left(\frac{4x}{4+x^2+y^2}, \frac{4y}{4+x^2+y^2}, \frac{2(x^2+y^2)}{4+x^2+y^2} \right) .$$

At (x,y) we can compute the partial derivatives:

$$f_x(x,y) = \frac{4}{(4+x^2+y^2)^2}(4 - x^2 + y^2, -2xy, 4x)$$

$$f_y(x,y) = \frac{4}{(4+x^2+y^2)^2}(-2xy, 4 + x^2 - y^2, 4y) .$$

It can be checked easily that these vectors are orthogonal and of the same length. Thus, near (x,y), the function f is indistinguishable from its differential, which preserves angles and ratios of lengths of vectors in the coordinate x and y directions. Since the differential is linear it preserves angles between all vectors.

Example of Students' Work on Problem 44

Professor and T.A. comments:

Student answer:

Problem 44 - Lucy Mapaseka Dladla

Suppose that the plane is tangent to the sphere at its South Pole S. The image of a parallel of latitude from a circle with center S and the image of a meridian circle form a straight line through S. The image of the equator is a circle on the plane, and the radius of the image circle is twice as large as the radius of the sphere.

Why? Explain

Which line through N ?

By parallel projection to the line through N and with the lines through the vectors, the angle formed by vectors at x is the same as the angle formed by image vectors or the plane on $S(x)$. Therefore, the stereographic projection preserves sizes of angles.

Response to comments:

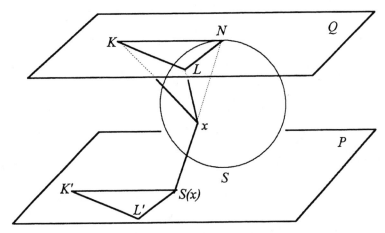

The angle between the two vectors a and b is defined as the angle between the tangents to these vectors at the point of their intersection x.

There are two vectors from the point x on the sphere. Let the tangents to these vectors in the point x intersect the plane Q, tangent to the sphere in the point N; in point L and K.

Now connect the points K and L with the point N. $Kx=KN$ as two tangents to the sphere drawn from the same point; and $Lx=LN$ due to the same reasons.

$\triangle KLx \cong \triangle KLN$ by SSS, therefore, $\angle KxL \cong \angle KNL$.

Now the vectors a and b are projected onto plane P as two vectors emerging from the point $S(x)$, the angle between these vectors being equal to that between the tangents. These tangents $S(x)K'$ and $S(x)L'$ are projections of the tangents xK and xL and are therefore, intersections of the planes NKx and NLx with the projection plane P. But the planes NKx and NLx intersect the plane Q, parallel to plane P, along the straight line NK and NL. Therefore the straight lines $S(x)K'$ and $S(x)L'$ are parallel, respectively to the lines NK and NL, $\angle K'S(x)L' \cong \angle KNL$; since $\angle KNL \cong \angle KxL$, we obtain $\angle K'S(x)L' \cong \angle KxL$. Therefore, $\angle KxL$ is preserved after projection to $\angle K'S(x)L'$.

Good.

Chapter 16

Duality and Trigonometry

In this chapter, we will first derive, geometrically, expressions for the circumference of a circle on a sphere, the Law of Cosines on the plane, and its analog on a sphere. Then we will talk about duality on a sphere. On a sphere, duality will enable us to derive other laws that will help our two-dimensional bug to compute sides and/or angles of a triangle given ASA, RLH, SSS, or AAA. Finally we will look at duality on the plane.

PROBLEM 45. *Circumference of a Circle.*

Find a simple formula for the circumference of a circle in terms of its intrinsic radius and make the formula as intrinsic as possible.

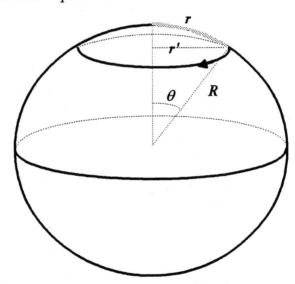

In the figure above, rotating the segment of length r' through a whole revolution produces the same circumference as rotating r, which is an arc of the great circle as well as the intrinsic radius of the circle on the sphere. The circumference of the circle can be given extrinsically by $C = 2\pi r'$, but since we know that $r' = R \sin \theta$ and $\theta = r/R$, we can also say that:

$$C = 2\pi R \sin(r/R).$$

Further, given that the length of a great circle, *g.c.*, is $2\pi R$, the expression above can be written in the following way:

$$C = g.c. \sin(2\pi r/g.c.).$$

All of the quantities given in this expression are intrinsic. Since the derivation of this expression involved extrinsic considerations, the reader may wish to ponder the question:

How could our 2-dimensional bug derive this formula?

By looking at very small circles, the bug could certainly find uses for the trigonometric functions that they give rise to. Then the bug could discover that the geodesics are actually circles, but circles which do not have the same trigonometric properties as very small circles. And then what?

Example of Students' Work on Problem 45

Professor and T.A. comments: *Student answers:*

Problem 45 - Andrew E. Sundstrom

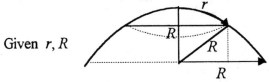

Given r, R

OK, but what is C for $r \neq 1/4(2\pi r)$?

As $r \to 1/4(2\pi R)$, $C_r \to 2\pi R$.
$C = r / (1/4(2\pi R)) \, 2\pi R = 4r$

Response to comments:

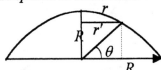

$r' = R \cos\theta$
$\theta = r/(1/4(2\pi R)) = 2r/\pi R$
$r' = R \cos(2r/\pi R)$
$C = 2\pi r' = 2\pi R \cos(2r/\pi R)$

Let us now derive the Law of Cosines, first on the plane and then on the sphere:

PROBLEM *46. Law of Cosines (on the Plane)*

If we know two sides and the included angle of a small triangle then, according to SAS, the third side is determined. If we know the lengths of the two sides and the measure of the included angle, how can we find the length of the third side?

Geometric Proof:

The following visual proof which is given in the students' manual comes from [**G:** Valens]:

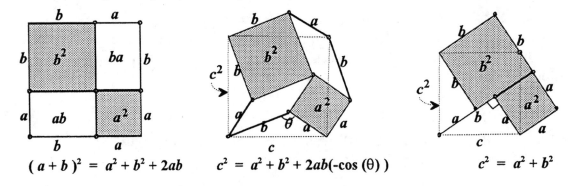

$(a+b)^2 = a^2 + b^2 + 2ab$ $c^2 = a^2 + b^2 + 2ab(-\cos(\theta))$ $c^2 = a^2 + b^2$

The proof for the case $\theta < \pi/2$ can be seen in the following picture:

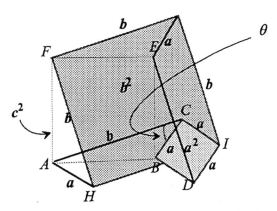

Note that here $\triangle ABC \cong \triangle FEG \cong \triangle EBD \cong \triangle EAH$. Then you can check that

$$[ABEF] = [BDIC] + [HIGF] - [AHIC] - [DIGE]$$

or

$$c^2 \ = \ a^2 \ + \ b^2 \ - \ 2\,[ab \cos \theta].$$

For a nice algebraic proof of the Law of Cosines on the plane the reader should see the following student example.

Example of Students' Work on Problem 46 (on the Plane)

Student answer:
Problem 46 (on the plane) - Jeffrey Cohen

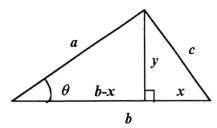

$$c^2 = x^2 + y^2$$
$$a^2 = y^2 + (b - x)^2 = x^2 + y^2 + b^2 - 2bx$$
$$y^2 = a^2 - b^2 - x^2 + 2bx$$
$$c^2 = a^2 - b^2 + 2bx$$
$$\cos\theta = \frac{b-x}{a}$$
$$a\cos\theta = b - x$$
$$x = b - a\cos\theta$$
$$c^2 = a^2 - b^2 + 2b(b - a\cos\theta) = a^2 + b^2 - 2ab\cos\theta \,.$$

Problem 46. Law of Cosines (on a Sphere)

To be able to find an expression for the Law of Cosines on a sphere, we suggest that the students use a particular gnomic projection. Students should pick a triangle, consider one of its vertices as the south pole of the sphere, and project the triangle from the center of the sphere. This gnomic projection will project two adjacent sides of the triangle onto straight lines on the image plane (the plane tangent to the vertex chosen as the south pole) and the angle between them will be preserved through this particular projection. We now use radians to measure the length of sides which is equivalent to taking the radius of the sphere, R, as the unit. Let us see how to derive:

$$\cos c = \cos a \cos b + \sin a \sin b \cos \theta.$$

Of course this expression is not the only one that students will derive.

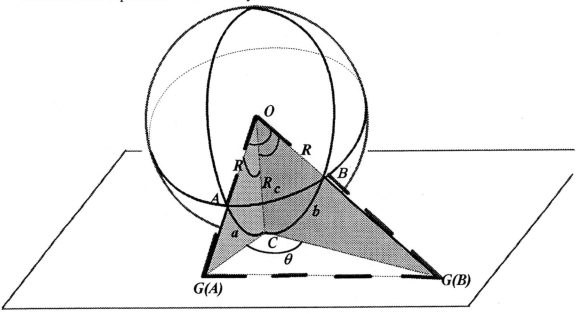

To better see the details of this picture, we look separately at four triangles:

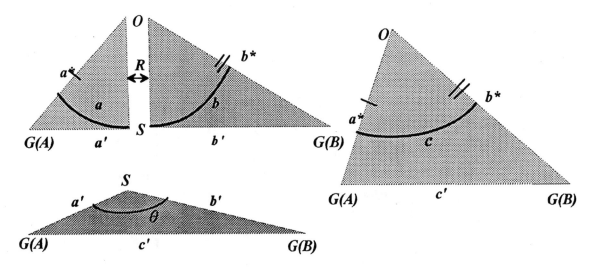

Using the four triangles shown above, we can obtain two expressions for the segment c' connecting $G(A)$ and $G(B)$, which is the projection of c:

$$c'^2 = a^{*2} + b^{*2} - 2\,a^*b^* \cos c = a'^2 + b'^2 - 2\,a'b' \cos \theta.$$

and

$$a' = a^* \sin a, \quad b' = b^* \sin b.$$

By algebraically manipulating these expressions, we obtain a law of cosines on a sphere:

$$\cos c = \cos a \cos b + \sin a \sin b \cos \theta.$$

Example of Students' Work on Problem 46 (on a Sphere)

The student's first attempt at this problem ended in pages of futile algebraic manipulations. Because of its length we decided not to include it here. She makes a brief mention of her first attempt and then proceeds with a new start.

Student answer:

Problem 46 on the sphere - Varuni Kondagunta

On the sphere, I guess my idea of making each plane tangent to the sphere at each vertex was kind of right, but I failed to make the 3-D connection between all three planes.

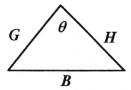

$$G = R \tan a, \quad x = R \sec a, \quad H = R \tan b, \quad y = R \sec b$$

$$B^2 = x^2 + y^2 - 2xy \cos c = R^2[(\sec^2 a + \sec^2 b) - 2 \sec a \sec b \cos c]$$

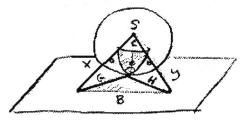

$$B^2 = G^2 + H^2 - 2GH \cos \theta = R^2[(\tan^2 a + \tan^2 b) - 2 \tan a \tan b \cos \theta]$$

$$\tan^2 a + \tan^2 b - 2 \tan a \tan b \cos \theta = \sec^2 a + \sec^2 b - 2 \sec a \sec b \cos c$$

but, $\sec^2 x = \tan^2 x + 1$; so, $\sec^2 x - \tan^2 x = 1$

$$(\sec^2 a - \tan^2 a) + (\sec^2 b - \tan^2 b) =$$

$$-2 \tan a \tan b \cos \theta + 2 \sec a \sec b \cos c = 2$$

$$-\tan a \tan b \cos \theta + \sec a \sec b \cos c = 1$$

$$\sec a \sec b \cos c = 1 + [\tan a \tan b \cos \theta]$$

$$\frac{\cos c}{\cos a \cos b} = 1 + \frac{\sin a \sin b \cos \theta}{\cos a \cos b}$$

$$\frac{\cos c}{\cos a \cos b} = \frac{\cos a \cos b}{\cos a \cos b} + \frac{\sin a \sin b \cos \theta}{\cos a \cos b}$$

$\cos c = \cos a \cos b + \sin a \sin b \cos \theta$
The Law of Cosines on sphere.

PROBLEM 47. Law of Sines

If $\triangle ABC$ is a triangle on the plane with sides, a, b, c, and corresponding opposite angles, α, β, γ, then

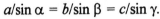

$$a/\sin \alpha = b/\sin \beta = c/\sin \gamma.$$

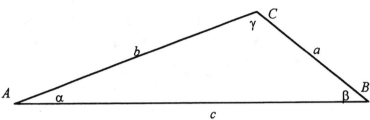

What is an analogous property on the sphere?

The standard proof for the Law of Sines is to drop a perpendicular from the vertex C to the side c and then to express the length of this perpendicular as both $b \sin \alpha$ and $a \sin \beta$. From this the result easily follows.

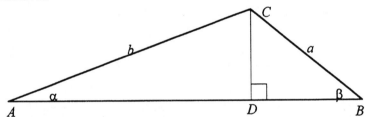

Thus, on the plane the Law of Sines follows from an expression for the sine of an angle in a right triangle. For triangles on the sphere, we can find a very similar result. If $\triangle ACD$ is a triangle on the sphere with the angle at D being a right angle, then use gnomic projection to project $\triangle ACD$ onto the plane which is tangent to the sphere at A. Since the plane is tangent to the sphere at A the size of the angle α is preserved under the projection. In general, angles on the sphere not at A will not be projected to angles of the same size, but in this case the right angle at D will be projected to a right angle.

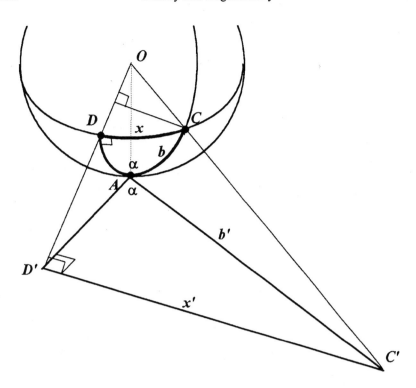

One can argue by symmetry that the angles $\angle AD'C'$ and $\angle OD'C'$ are right angles and already $\angle OAC'$ is also a right angle. The students will probably need to draw careful pictures and make models to see this. Now $\sin \alpha = x'/b'$ and, we can see that, $b' = \tan b$ and $x' = (\sin x)(\sec b)$. We conclude that

$$\sin \alpha = (\sin x)(\sec b)/(\tan b) = \sin x/\sin b.$$

Thus, for a right triangle on a sphere, 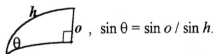 , $\sin \theta = \sin o / \sin h$.

Putting this together (as in the plane proof) we get, for an arbitrary (small) spherical triangle:

$$\sin a / \sin \alpha = \sin b / \sin \beta = \sin c / \sin \gamma.$$

The expressions in the Law of Cosines above give us the measure of the unknown side of a SAS triangle. We can now ask, for any version of ASA, ITT, SSS, ASS, SAA, or AAA that we proved to be true on the sphere, can we also find expressions for those missing sides or angles? To answer this, we introduce the notion of duality here. When solving Problems 6 and 7, students probably noticed that there is a duality between lines and points, that is, in SAS we have that two points determine a unique line and in ASA we have that two intersecting lines determine an unique point. This fact is a good motivation for introducing the notion of duality on a sphere.

Problem 48 asks students to find the dual of a triangle (a small triangle) and to find the relationships between the sides, angles and vertices of the original and the dual triangles.

PROBLEM 48. *The Dual of a Small Triangle*

Find the relationship between the sizes of the angles and sides of a triangle and the corresponding sides and angles of its dual.

The dual of a triangle is another triangle, $\Delta A^*B^*C^*$, whose sides are the duals of the exterior angles of the original triangle and whose exterior angles are the duals of the original triangle's sides. Pictorially:

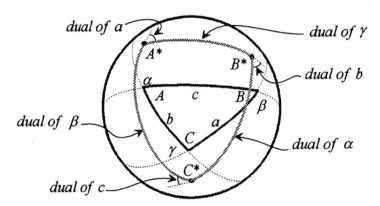

Note that points inside the original triangle correspond to lines in the exterior of the dual triangle. If we take a triple right triangle, then its dual coincides with itself. Note that, throughout the book, we have been bumping into the exterior angles of a triangle and in a way that shows that the exterior angles are oftentimes more relevant for expressing properties of triangles than the interior angles.

PROBLEM 49. Trigonometry on Spherical Triangles.

In each of ASA, RLH, SSS, AAA, if you know the measures of the given sides and angles, how can you find the measures of the sides and angles that are not given?

Note: the notations we will use in Problem 49 refer always to the picture in Problem 48:

ASA:

Working with small triangles assume that you know angles A and C and the included side b. Then, from the dual triangle, we know sides α and γ and the included angle $(\pi - b)$. We are then in a SAS situation that can be solved using the Law of Cosines, that is:

$$\cos\beta = \cos\alpha\cos\gamma + \sin\alpha\sin\gamma\cos(\pi - b).$$

Now, $\alpha = \pi{-}A$, $\gamma = \pi{-}C$, $\beta = \pi{-}B$, and, for any angle θ, $\cos(\pi - \theta) = -\cos\theta$, $\sin(\pi\square - \theta) = \sin\theta$. Thus, the above expression becomes:

$$-\cos B = \cos A\cos C - \sin A\sin C\cos b.$$

We were able to determine the third side of the dual triangle. Using the law of cosines two more times, we can determine the three sides of the original triangle through the following formulas:

$$\cos a = \frac{\cos A + \cos C\cos B}{\sin C\sin B} \text{ and } \cos c = \frac{\cos C - \cos B\cos A}{\sin B\sin A}.$$

SSS or AAA:

The case SSS is the dual of the case AAA; that is, if we can determine angles then on the dual we will be able to compute the sides. Let us find the algebraic expressions for the exterior angles given the sides. If the sides a, b and c are given, then we have:

$$\cos A = \frac{\cos a - \cos b \cos c}{\sin b \sin c}, \quad \cos B = \frac{\cos b - \cos c \cos a}{\sin c \sin a}, \quad \text{and } \cos C = \frac{\cos c - \cos a \cos b}{\sin a \sin b}.$$

If we are given the three angles A, B and C, then we take the duals of the equations above:

$$\cos a = \frac{\cos A + \cos B \cos C}{\sin B \sin C}, \quad \cos b = \frac{\cos B + \cos C \cos A}{\sin C \sin A}, \quad \text{and } \cos c = \frac{\cos C - \cos A \cos B}{\sin A \sin B}.$$

RLH:

   If we have a right triangle  , then the Law of Cosines gives us:

$$\cos c = \cos a \cos b.$$

   This case is like a Pythagorean theorem for the sphere. Now, if we are given sides b and c, then we can solve for side a if $\cos b \neq 0$ or $b \neq \pi/2$ (1/4 great circle).

SAA and ASS:

   For SAA, suppose we know the side a and the angles B and A, then by the Law of Sines for the sphere, $\sin b = (\sin B)(\sin a)/(\sin A)$ which has two solutions but only one of the solutions is less than a quarter great circle. (Similarly, for ASS we can use the Law of Sines to get a unique second angle less than $\pi/2$.) The third side can then be calculated by using the Law of Cosines but the result is very complicated. However, the formula for RLH can be translated into a formula for SAA in the special case that the given side is $\pi/2$ or 1/4 great circle.

Example of Students' Work on Problem 49

Student answer:

Frank Sinnock - Problem 49

ASA triangle:

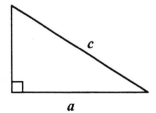

   The dual of the above triangle gives a triangle SAS from which we can get the third side with the Law of Cosines which is actually the third angle. Now, in the dual, by switching which side are a, b and c in the Law of Cosines formula, you can get the other two angles which are actually the other two sides of the original triangle.

   RLH triangle:

The last side can be found by using 46 [Law of Cosines], but solving for b instead of c. Then, switching the sides in the formula, you can get the other two angles.

SSS triangle:

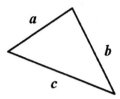

Solve for the angle in 46. Switch where the sides are put in the formula to get the other angles.

AAA triangle:

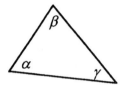

PROBLEM 50. Properties on the Projective Plane

Check that the following properties hold on the projective plane:

a. Two points determine a unique line.

b. Two parallel lines share the same point at infinity.

c. Two lines determine a unique point.

d. If a point is on a line, then the dual of the line is a point which is on the dual of the original point.

e. If C is the circle with center at the origin and radius the same as the radius of the sphere, then the dual of a point on C is a line tangent to C.

Proofs:

a. If the two points are not on the line at infinity then this is the usual property of the plane. But, even if one or both points is on the line at infinity, two different points on the projective plane are images of two different point pairs points on the sphere. These two different point pairs on the sphere define a unique great circle. The image of this great circle is a unique straight line on the projective plane that goes through the points we started with. If both points are on the line at infinity then they determine the line at infinity. If only one point is on the line at infinity, then the line determined is the line through the finite point in the direction of the infinite point.

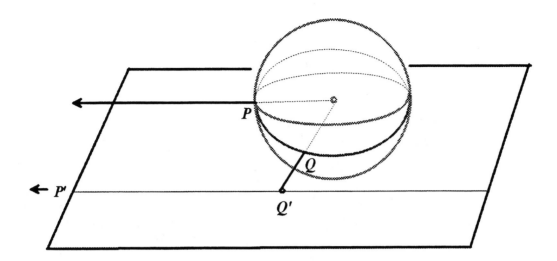

b. The two parallel lines are images of two great circles on the sphere. Imagine that the great circles intersect at a point Q not on the equator. The image of this point would be a point Q' on the projective plane, that is, on the two lines we started with. But, then the lines would intersect, which is not possible given that they are parallel lines. Thus the two great circles must intersect at a point-pair P on the equator. This point-pair is projected onto a unique point P' at infinity.

c. This is an ordinary property of non-parallel lines on the plane. Two lines on the projective plane are images of two great circles on the sphere. These great circles intersect in one point-pair which is projected onto a unique point in the projective plane. If the two lines are parallel then by what we have seen in part b., they determine the same point at infinity.

d. Let P be a point on the line l in the projective plane. If Q gets projected onto P, then by the definition of gnomic projection, Q lies on a great circle g which is projected onto l. Consequently, the dual of g is going to be a point that is the pole of g, that is, a point that is at 1/4 great circle distance from every point on g. Given that Q is a point on g and Q is 1/4 great distance from every point of its dual then h, the dual of Q, has to go through the dual of g. Projecting h into the plane we have that the dual of l (the projection of the dual of g) is on the dual of P (the projection of the dual of Q).

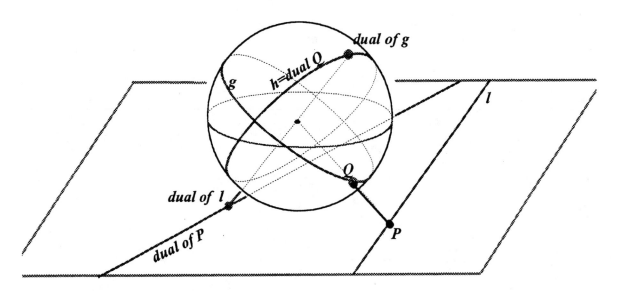

e. A pictorial proof:

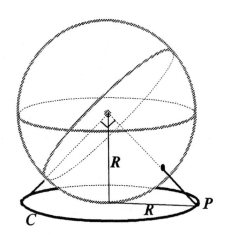

Chapter 17

Isometries and Patterns

In this chapter, students will look at some fundamental concepts of group theory through its geometric origins. Students have a notion of what a symmetry is since Chapter I, however, embedding the concept in group theory by exploring patterns is not easy for students. In the students manual, we try to make the terminology simple and explain what it means through concrete examples.

PROBLEM 51. *Examples of Patterns*

Find as many (non-isomorphic) patterns as you can which have only finitely many symmetries.

Find as many (non-isomorphic) strip patterns as you can.

List the symmetries of each.

In Problem 51, students are asked to look out at the world and find finite patterns and strip patterns. Old buildings, places of worship, decorative china and wallpaper edging all have a diverse variety of patterns. Ask students to find as many as they can. The purpose of this problem is not for the students to find a complete list, but rather for them to start focusing on patterns and their symmetries.

PROBLEM 52. *Isometry Determined by its Action on Three Points*

Prove the following: On the plane or sphere, if f and g are isometries and A, B, C are three non-collinear points, such that $f(A) = g(A)$, $f(B) = g(B)$, and $f(C) = g(C)$, then f and g are the same isometry, that is, $f(X) = g(X)$ for every point X.

Linear Algebra Proof:

Three non-colinear points, A, B, and C, determine a plane. You can pick one of the points (say C) as the origin and consider the other two as vectors, A, B, and then any point (vector) X in the plane is a linear combination of A and B. Any two linear functions that coincide on a basis will coincide on every point X. In general, an isometry, f, is not a linear; however, the function g defined by $g(X) = f(X) - f(C)$ can be shown to be always linear and thus Problem 52 follows.

The application of linear algebra seems awkward when applied to isometries. Affine linear algebra is less awkward, but unfortunately few students have even heard of it. In any case, we have not seen any proofs from students that proceed from linear algebra whether affine or not. Students prefer to show directly that the images of three non-collinear points determine the isometry on every point in the plane. Intuitively, three points determine a plane and an isometry moves the whole plane in a rigid way. The instructor should expect proofs from the students with different levels of sophistication.

Here is a direct proof from a student that is illustrative:

Example of Students' Work on Problem 52

Professor and T.A. comments:

Student answer:

Problem 52 - Robert Pless

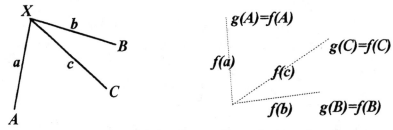

OK!

Lengths are maintained through transformations, so lengths:

$a = f(a) = g(a)$; $b = f(b) = g(b)$; $c = f(c) = g(c)$. I am not yet saying that $f(X) = g(X)$ or $f(a)$ is the same line as $g(a)$, just the lengths are the same!

So we have three circles, centered at A, B, C of radii, a, b, c, whose intersections are all possible locations of $f(X)$ and $g(X)$.

The intersection of 3 circles:

Consider the intersection of 2 circles: at most there are 2 intersection points:

Nice.

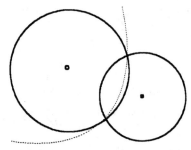

Any third circle q that intersects both points must have a center that is collinear with the centers of the other circles -- because they all share a common chord and the center of a circle is on the perpendicular bisector of any chord of a circle by Lemma 1 [on next page].

We know our circles are non-collinear, because we chose them that way so the circles met in only one point. So $f(X)$ must be that point and $g(X)$ must be that point, so $f(X) = g(X)$.

LEMMA 1:

The center of a circle lies on the perpendicular bisector of a chord.

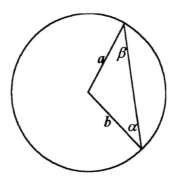

$a = b$ because they are both equal to radius of the circle; $\alpha \cong \beta$ by ITT, so we have: draw l from the center such that it intersects at rt. angles with the chord:

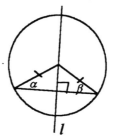

On the plane if we know two angles we know the third, so the triangles are congruent by ASA. Thus l is also the bisector of the chord.

So the center of the circle lies on the perpendicular bisector of the chord.

Notice that the student's Lemma 1 was proved previously as a direct corollary to ITT for both the plane and the sphere. See Chapter 7.

PROBLEM 53. Classifications of Isometries on the Plane and Sphere

To solve Problem 53 we will use Problem 52. That is, in each case we will choose three non-collinear arbitrary points (or more simply a triangle) and prove the statement for those points. Then by Problem 52, the isometries in question will coincide at every point.

a) Every isometry is the composition of one, two or three reflections:

Take a triangle on the plane and transform its vertices through an isometry. This is the same as considering any two congruent triangles sitting on the plane. Look at the following reflections pictured below:

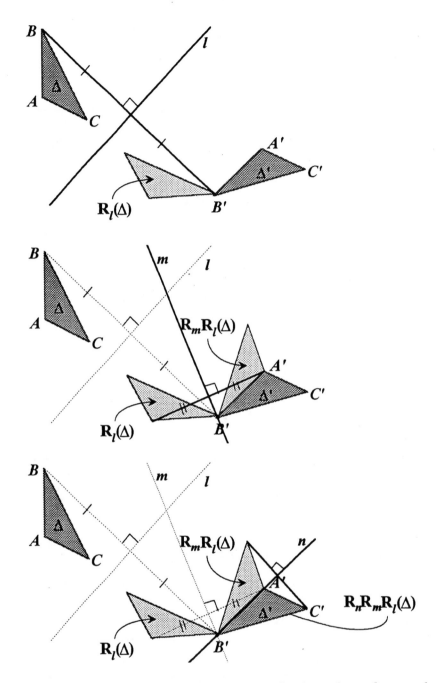

By Problem 52, the isometry is the composition of the three reflections about l, m, and n.

b) Every composition of two reflections is either a translation, a rotation or the identity. How can you tell which one?

Take any two lines on the plane and consider their relative positions:

Case 1: The lines of reflection do not coincide and are parallel.

Let the distance between the two lines be d and consider the action of the two reflections. Since Problem 52 allows us to choose any three non-collinear points, we can choose them conveniently as indicated:

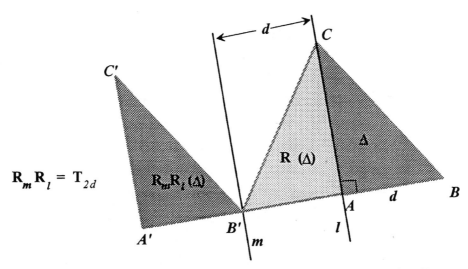

Note that the composition of these two reflections is a translation through a distance $2d$ in the direction perpendicular to the parallel lines of reflection. Thus, this same translation is also the composition of two reflections about any pair of parallel lines which are a distance d apart and which are perpendicular to the direction of the translation. Also note that if the reflections are composed in the other order, then the translation goes in the reverse direction.

Case 2: The lines of reflection meet at a point P and the angle between them is θ.

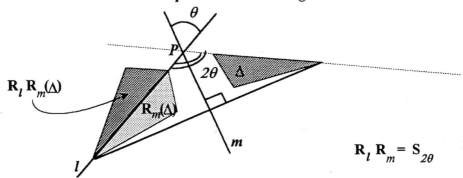

Using a theorem that we proved back in Chapter 7, we know that perpendicular to the base of an isosceles triangle is the bisector of the angle opposite the base. We can, therefore, conclude that $R_m R_l = S_{2\theta}$.

Note the same rotation is the composition of any two reflections about lines which intersect at P in an angle of θ.

c) Every composition of three reflections is either a reflection or a glide reflection. How can you tell which one?

Take three reflection lines and consider their relative positions.

Case 1: The three lines are parallel to each other.

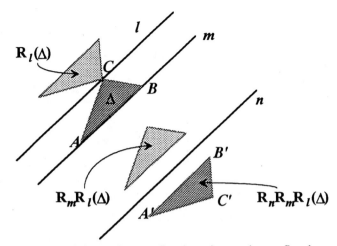

We can see that the composition of these three reflections is another reflection.

Case 2: The three lines intersect at a point P.

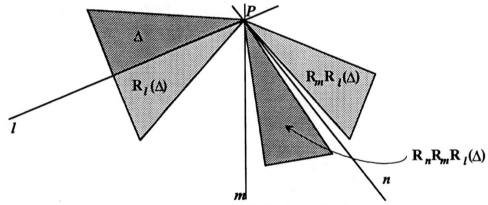

Again we can see that the composition $R_n R_m R_l$ is a reflection.

Case 3: The first two lines are parallel and the third is a transversal of both.

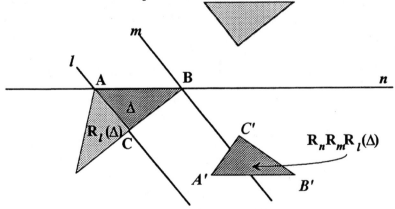

We can see that the composition of the three reflection is a glide reflection along a line parallel to n.

Case 4: The three lines intersect in three points.

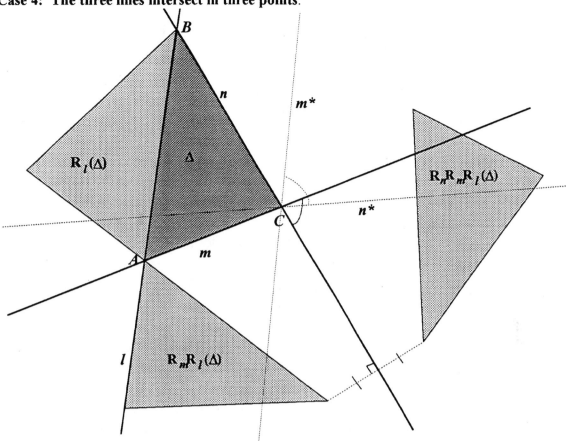

Note that $R_n R_m R_l = (R_n R_m)R_l$. By part b), $R_n R_m$ is a rotation about C, the intersection of m and n; and this rotation is also equal to the composition of $R_{n*} R_{m*}$ where m^* and n^* are any lines through C making the same angle as m and n. Now choose m^* and n^* so that m^* is parallel to l. Then

$$R_n R_m R_l = (R_n R_m)R_l = (R_{n*} R_{m*})R_l = R_{n*} R_{m*} R_l$$

which satisfies Case 3 and thus $R_n R_m R_l$ is a glide reflection along a line parallel to n^*.

Another proof of c):

In the figure below we started with the triangles $\triangle ABC$ and $\triangle A'B'C'$, where $\triangle A'B'C'$ is image of $\triangle ABC$ through $R_n R_m R_l$. Prolong two corresponding sides of the congruent triangles and consider the bisector, b, of the angle between the two lines. Parallel transport this line (along one of the sides that we lengthened) to the line, p, that is equidistant from A and A'. Then we have that:

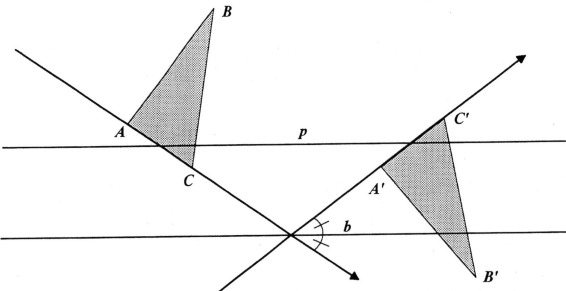

Then we see that $R_n R_m R_l$ is equal to a glide refection along p or a reflection through p. This argument can be used to give a direct proof of d), outlined below.

   d) Every isometry of the plane or sphere is either a reflection, a translation, a rotation, a glide reflection, or the identity.

   The following is an outline of a direct proof of d):

- If f is an isometry and $\triangle ABC$ is any triangle, then the triangle is congruent to $\triangle A'B'C'$, where $f(A) = A'$, $f(B) = B'$, $f(C) = C'$. This congruence is either a direct congruence (no reflection needed) or not a direct congruence.

- If the congruence is direct and the lines AA', BB', CC' are parallel then the isometry is a translation along the direction of AA'.

- If the congruence is direct and the lines AA', BB', CC' are not parallel then the isometry is a rotation about the intersection of the perpendicular bisectors of the lines AA', BB', CC'.

- If the congruence is not direct then the **Another proof of c)** directly above will show that the isometry is a glide reflection or reflection.

Example of Students' Work on Problem 53

   Professor and T.A. comments: Student answer:

Problem 53 - Brian Smits
Every isometry is a composition of 3 or fewer reflections.
   Proof: Let f be an isometry. Let ABC be a triangle and $A'B'C' = f(ABC)$ we now construct an isometry f composed of three or fewer reflections that also maps ABC to $A'B'C'$ and is therefore equivalent to f. With one reflection, we can map A to A' by reflecting across the perpendicular bisector of the line containing A

Why?

and A'. We apply that reflection to all points. Now we need to map B to B'. We can do this by reflecting across the angle bisector of $B''A''B'$ ($=B''A'B'$). Since the reflection was across a line through A', the reflection has no affect on the A's. Now A and B line up, and because the triangle is rigid, C''' can be aligned with C' if needed by reflecting across the line between A' and B'.

There are three possibilities for two reflections. Either the two lines cross, they are parallel and disjoint, or they are the same line. If the lines are the same the two reflections cancel each other, creating the identity. If the two lines cross then exactly one point remains fixed.

Since the two reflections result in an unreflected image, the only possible rigid transformation that meets these constrains is a rotation. If the two lines are parallel and disjoint, then no point remains fixed. Also, the two reflections map infinite perpendicular transversals to themselves (including orientation). The only rigid transform that maps an infinite line onto itself but keeps no point fixed is a translation.

Are you sure? Why?

This next statement is wrong. There are several possibilities for three reflections. Three parallel reflections create a translation/reflection but not a glide reflection because the reflection is not across the line of translation. If the three lines cross in different places, then any two of then form a rotation, but this is not a glide reflection either. The easiest way to get three reflections to produce a glide reflection is to have two parallel lines cut by a transversal at an angle of $\pi/2$.

but given 3 lines of reflection intersecting in 3 points, how can you prove that it is the same as a glide?

The only rigid transformations are reflections, translations or rotations. A glide reflection is the composition of two rigid transformations. Since an isometry must be a rigid transformation, every isometry is one of these.

Not clear.

Response to comments:

Why? Say more about how A is picked..

Let the first two lines of reflection cross at an angle of θ at point B. The two lines produce a rotation of 2θ around B. Now we can look at two points A and B and see what the rotation plus reflection does to them. B remains stationary through the rotation and A gets rotated just far enough so that it remains on the line of reflection. The two isosceles triangles create a parallelagon. Looking at the parallelagon, it is easy to find a glide reflection that maps A and B to A' and B'. It takes three points to define an isometry, but the proof to show that two isometries are the same only used the third point to remove the ambiguity that a reflection might cause after transforming the first two points. Here we can determine directly that the third points will line up because there are the same number of reflections involved (a translation is equivalent to two reflections and a glide reflection is a translation plus a reflection).

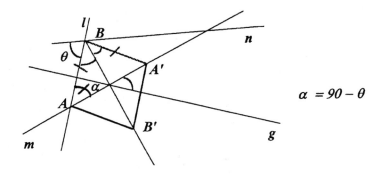

$$\alpha = 90 - \theta$$

Response to comments:

    A gets picked so that after the rotation it remains on the third line of reflection. This means that the angle at *B* is 2θ and the lines from *B* to *A* and *B* to *A'* form an isosceles triangle with the third reflection line. I explained this in class a bit more thoroughly than this.

Problem 54. *Classification of Discrete Strip Patterns*

Prove there are only seven strip patterns on the plane which are discrete.

What are some non-discrete strip patterns?

What happens with strip patterns on a sphere?

    In Problem 53, we saw that an isometry is either a translation, a reflection, a rotation, a glide reflection, or the identity. By definition, a symmetry of a strip pattern is also a symmetry of a straight line along which there is translation symmetry of the pattern. Consequently, the only rotation symmetry of a strip pattern is the half-turn rotation symmetry, because it is the only rotation that takes a straight line onto itself. There are two reflections that take the line onto the line: reflection through the line and reflection perpendicular to the line. The only glide reflection symmetry possible is a glide along the line. We present now a list of all strip patterns, seven in total:

Translation:

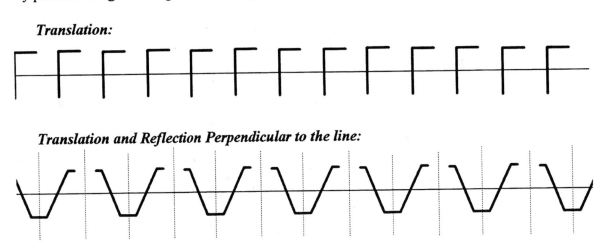

Translation and Reflection Perpendicular to the line:

Translation and Reflection through the line:

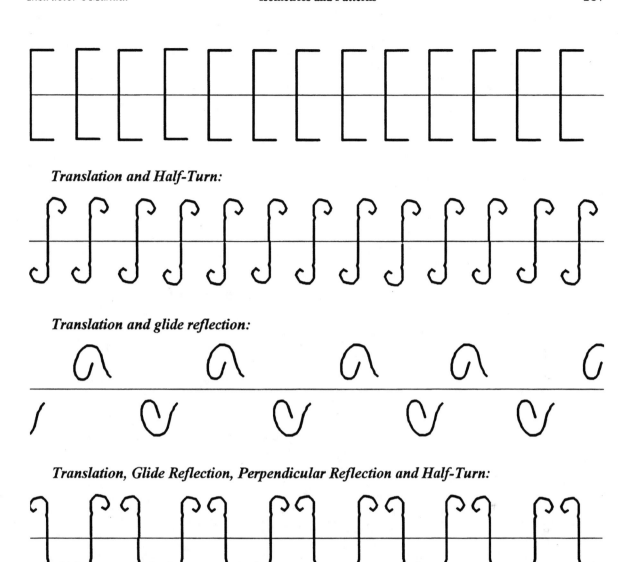

Translation and Half-Turn:

Translation and glide reflection:

Translation, Glide Reflection, Perpendicular Reflection and Half-Turn:

Translation, Reflection, Perpendicular Reflection, Glide Reflection and Half-turn:

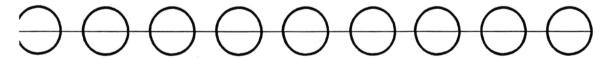

Strip Patterns on the Sphere.

The above seven patterns also occur on a sphere except that the patterns wrap around the sphere and, therefore, come back onto themselves. A strip pattern is discrete if the smallest translation symmetry is through a distance which is commensurate with the length of a great circle. For each of the combinations of symmetries above, there is a different strip pattern for each positive integer (the number of motif in the pattern along the great circle).

Example of Students' Work on Problem 54

Professor and T.A. comments:

Student answer:

Problem 54 - Frank Jary

I am going to prove the existence of only 7 discrete strip patterns.

First, we must keep in mind that the only symmetries we have are made of the following:

$R_\perp$ - Reflection about any perpendicular line of a line l

R_l - Reflection about a line l

H - Half-turn symmetry which is a composition of $R_\perp$ and R_l.

T - Translation along l, which is a composition of $R_\perp R_\perp$

G - Glide along l which is a composition of R_l and T

We can generate all possible combinations of these (T is part of all of them, and I will, therefore, leave it out)

$R_\perp$, $R_\perp R_l$, $R_\perp H$, $R_\perp G$,

$R_\perp R_l H$, $R_\perp R_l G$, $R_\perp H R_l$, $R_\perp HG$, $R_\perp G R_l$, $R_\perp GH$

(These six all generate $R_\perp R_l HG$)

R_l, $R_l R_\perp$, $R_l H$, $R_l G$, $R_l R_\perp H$, $R_l R_\perp G$, $R_l H R_\perp$, $R_l HG$, $R_l G R_\perp$, $R_l GH$

(The last six all generate $R_\perp R_l HG$)

H, $HR_\perp$, HR_l, HG, $HR_\perp R_l$, $HR_\perp G$, $HR_l R_\perp$, $HR_l G$, HGR_l, $HGR_\perp$

(The last six generate $R_l R_\perp HG$.)

G, $GR_\perp$, GR_l, GH, $GR_\perp R_l$, $HR_\perp G$, $GR_l R_\perp$, $GR_l H$, GHR_l, $GHR_\perp$

(The last six generate $R_l R_\perp HG$.)

T is also a possibility.

Let's eliminate doubles:

T	$HR_\perp$	$R_\perp R_l H$	$R_\perp R_l G$
H	HR_l	$R_\perp R_l G$	
G	HG	$R_\perp HG$	
$R_\perp$	$GR_\perp$	$R_l HG$	
R_l	GR_l		
	$R_\perp R_l$		

Note: All of the above except T itself also have T.

$H = R_\perp R_l$, $T = R_\perp R_\perp$, $G = R_l T$,

R_l entails GR_l

$R_\perp R_i$ entails $R_\perp R_i H$, $R_\perp R_i G$, $R_\perp R_i HG$, H, G and HG
HR_i entails $R_i HG$
$R_\perp$ entails T
The only ones left are:
$R_\perp$; R_i; $HR_\perp$; HR_i; $GR_\perp$; $R_\perp R_i$; $R_\perp HG$

Nice.

We know that these are the only ones, because they must all be isometries or combinations of isometries. Thus, we have all combinations of reflection, translation, rotation. Rotation is half-turn symmetry in this case, and glide is a combination of a reflection and a translation.

Only the symmetries I have come out of the isometries. I checked that all other combinations are still variations on the five I have. We tested how each of translation and rotation came out of 53. A combination of:

Reflection and Translation $\Rightarrow$ Reflection
Reflection and Rotation $\Rightarrow$ 1). Glide or 2). Reflection (see diagram)
Translation and Rotation $\Rightarrow$ Rotation

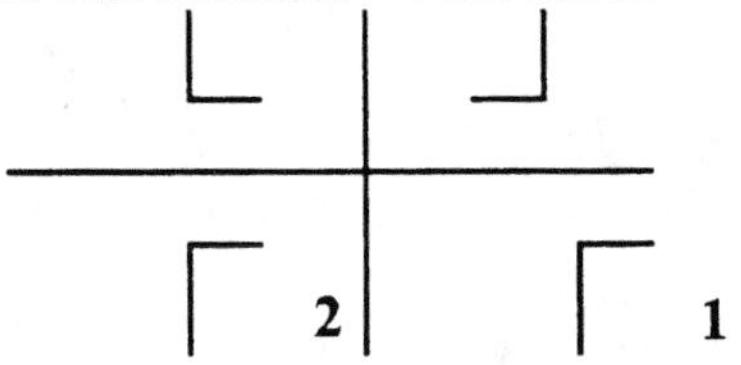

To be a strip pattern, by definition it must have a T. (But not necessarily a smallest.)

Non-discrete strip patterns are: (These do not have T)

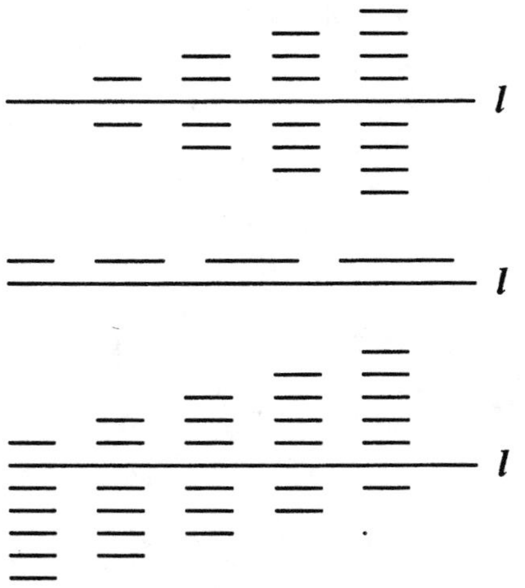

What happens on a sphere?
Depending on how the patterns are spaced and on their size, they may superimpose as they wrap around the sphere or they may not. The pattern will go around on a great circle, and if the multiple is any division that will result in an integer when divided into 360, the pattern

will be superimposing. Everything else will not be. If the patterns are very large (larger than 1/2 great circle), then the patterns will overlap.

Response to comments:
There's another non-discrete strip pattern. (Actually, the only one I could come up with immediately.)

l
———————————————————————————————————

———————————————————————————————————

PROBLEM 55. *Classification of Finite Plane Patterns*

If the group of symmetries of a geometric figure on the plane is finite, then translation and glide reflection isometries cannot be part of the symmetries of that geometric figure. Note that in all patterns laid out in Problem 54, if a translation by a is a symmetry of the pattern, then translations by $2a$, $3a$, $4a$,... are also symmetries of the pattern, that is, having a translation implies having an infinite number of translations and the group of symmetries is not finite. As a consequence, the only symmetries of a finite plane pattern are rotations and/or reflections. As we have seen in Problem 53, a rotation is the composition of two reflections. Now why is it that all the reflection lines intersect in only one point P? If there is a line of reflection that does not go through P, then we can construct a glide reflection as in Problem 53. If we have a glide reflection, we have an infinite number of translations and the pattern is not finite. The same argument proves that the center of symmetry is unique.

Note, however, that on a sphere the pattern can be finite without having only one center of symmetry. This is due to the fact that on a sphere rotations have two centers and a translation symmetry can exist without producing an infinite number of symmetries. In the figure below, the points A, B, C, D, E and F are all centers of symmetry of the pattern.

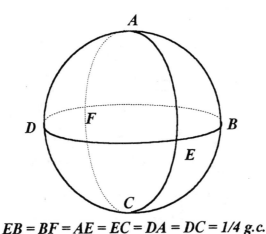

$$EB = BF = AE = EC = DA = DC = 1/4 \ g.c.$$

Examples of Students' Work on Problem 55

Professor and T.A. comments:

Student answer:

Why?

Problem 55 - Michael Kraizman

Any pattern in the plane with only finitely many symmetries has a center. That is, $f(p) = p$ for all f an isometries.

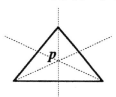

This point p, is located at the intersection of all the symmetries of the figure (see above). Since this point is located on all the symmetries and since symmetries take points along themselves to themselves, this point must stay the same regardless of which symmetry of the figure we consider.

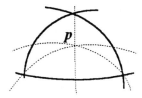

Why?

This is also true on the sphere. Again, this point is located at the intersection of all the symmetries of the figure. As long as we rotate about this point (also true in the plane), this point will still map to itself.

Describe all patterns on the plane with only finitely many symmetries:

A pattern with finitely many symmetries in the plane is any figure that is generated by a finite number of reflections or rotations or both, applied to a symmetric or asymmetric figure in the plane.

There are more patterns. (Some without reflections)

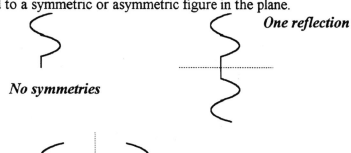

No symmetries

One reflection

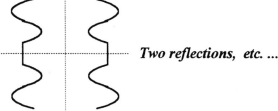

Two reflections, etc. ...

Why is the intersec-tion of all the symmetry lines unique?

Response to comments:

Why does any pattern in the plane with only finitely many symmetries have a center?

Since all symmetries (isometries) take a figure onto itself, if we draw in all the lines of symmetry for the figure, the point of intersection of all the symmetry lines is the center. That is, this is the point that does not get mapped to anything but itself during any isometry operation. Now, you might say, what about something like:

This figure only has one symmetry line. Where is its center? This is the only kind of special case that I can think of, and the answer is that the center of this figure is the line of symmetry (the whole line), since all points on this line will map to themselves under isometry transformations. For figures with multiple symmetry lines, the center is at the intersection of the symmetry lines.

Response to comments:

Why is the center unique?

The center is unique because we only have a finite number of symmetries. Since this is so, we cannot have symmetries such as glide, since that would move the figure off of itself (glide has a translation in it). So, why is it unique? A center, as I define it, is a point (or in the case of only one symmetry, a line) that maps onto itself under all isometries. Suppose you had two centers in a figure. That is, let p and p' lie on the symmetries of the figure. Let p' be at the intersection of all the symmetry lines and p be on at least one symmetry different from p':

Why not

?

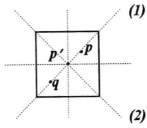

(1)

(2)

Then under transformation (1), p and p' are mapped to themselves. This satisfies the definition of center, but under transformation (2), p is mapped to q while only p' stays the same. So, p' has to be unique center because it is located at the point of null change.

Chapter 18

Polyhedra

In this chapter, students will prove theorems about tetrahedra, which are the 3-space analogs of planar triangles. In Problems 56-59, students will have the opportunity to use the theorems about triangles on the sphere that they proved in Chapters 5 through 9 as they investigate properties of tetrahedra. In Problem 60 the students will examine the regular solids.

PROBLEM 56. *Measure of a Solid Angle*

The measure of a solid angle is defined as the ratio,

$$m(\angle A) = [\lim_{R \to 0}] \frac{\text{area}\{ (\text{interior of } \Delta ABCD) \cap S \}}{R^2},$$

where S is any small 2-sphere with center at A whose radius, R, is smaller than the distance from A to each of the other vertices.

Show that the measures of the solid and dihedral angles of a tetrahedron satisfy the following relationship:

$$m(\angle A) = m(\angle AB) + m(\angle AC) + m(\angle AD) - \pi.$$

Show that two solid angles with the same measure are not necessarily congruent.

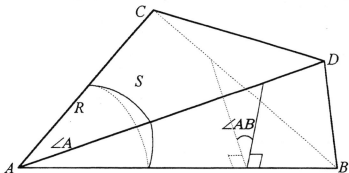

Proof:

The angles of the spherical triangle formed by $\{(\text{interior of } \Delta ABCD) \cap S\}$ can be seen to be the 3 dihedral angles of the tetrahedron around the vertex A:

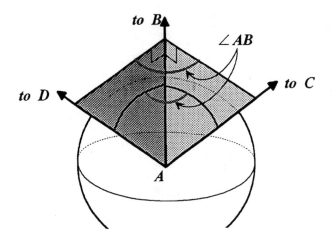

and then:

$$\{(\text{interior of } \triangle ABCD) \cap S\} = R^2 [m(\angle AB) + m(\angle AC) + m(\angle AD) - \pi].$$

Dividing by R^2 we get the desired result.

It is important to note that two solid angles can have the same measure without being congruent, given that two triangles on the sphere can have the same area without being congruent.

Examples of Students' Work on Problem 56

Student answer:

Problem 56 - Jun Kawashima

If we take a look at the cross-section for a dihedral angle and an angle on the spherical triangle,

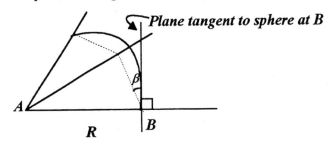

$\angle AB$ is congruent to the spherical angle β, and similarly:

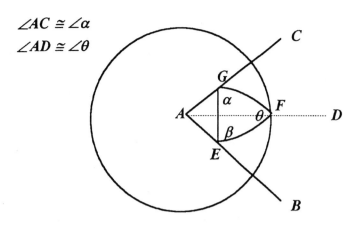

$$\angle AC \cong \angle \alpha$$
$$\angle AD \cong \angle \theta$$

$$m(\angle A) = \lim_{R \to 0} \frac{\text{area}\{(\text{interior of } \triangle ABCD) \cap S\}}{R^2}$$

$$= \lim_{R \to 0} \frac{\text{area(spherical triangle } GEF)}{R^2}$$

But from Problem 8 we have:

$$area(\Delta) = R^2 (\Sigma \angle's - \pi)$$

then, $m(\angle A) = \lim_{R \to 0} \dfrac{R^2(\alpha + \beta + \theta - \pi)}{R^2} = \alpha + \beta + \theta - \pi$

$$\Rightarrow m(\angle A) = m(\angle AB) + m(\angle AC) + m(\angle AD) -$$

Also, because there is an infinite number of combinations of $m(\angle AB) + m(\angle AC) + m(\angle AD)$ that sum to the same angle measure, solid angles with the same measure are not necessarily congruent.

PROBLEM 57. Edges and Face Angles

If $\triangle ABCD$ and $\triangle A'B'C'D'$ are two tetrahedra such that $\angle BAC \cong \angle B'A'C'$, $\angle CAD \cong \angle C'A'D'$, $\angle BAD \cong \angle B'A'D'$, $CA \cong C'A'$, $BA \cong B'A'$, $DA \cong D'A'$ then $\triangle ABCD \cong \triangle A'B'C'D'$.

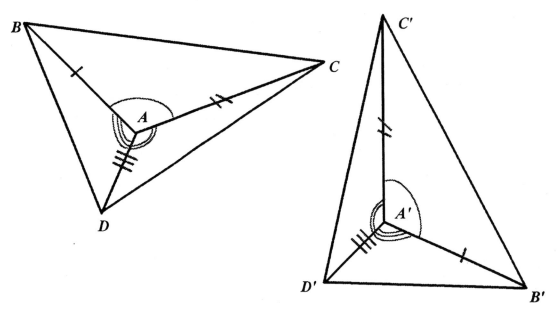

The spherical triangles corresponding to the solid angles A and A' have corresponding sides congruent given that they are the same as the face angles at A (A') (using radian measure). Consequently, the corresponding spherical triangles are congruent by SSS and then the solid angles A and A' are congruent. Now using SAS for the three triangle faces with one vertex at A (A') the corresponding faces of the tetrahedra at A (A') are congruent and using SSS for the bottom triangle face we see that all the corresponding faces of the tetrahedra are congruent. Thus, all the corresponding face angles at the vertices B (B') and C (C') are congruent and we can follow the same steps we used for the solid angles A and A'. Then, all the corresponding faces and corresponding solid angles are congruent; so, the tetrahedra are congruent.

Note that the work we did for Problem 57 allows us to conclude that if we have the spherical angle determined, for a solid angle A, the face angles with vertex A are also known as the dihedral angles at vertex A. This result will be used in Problem 58.

Examples of Students' Work on Problem 57

Student answer:

Problem 57 - Jason Fromberg

Since $S \cap$ (interior $\triangle ABCD$) has sides $R \times \angle BAC$, $R \times \angle CAD$, and $R \times \angle BAD$, and $S \cap$ (interior $\triangle A'B'C'D'$) has sides $R \times \angle B'A'C'$, $R \times \angle C'A'D'$, and $R \times \angle B'A'D'$, and since $\angle BAC \cong \angle B'A'C'$, $\angle CAD \cong \angle C'A'D'$, and $\angle BAD \cong \angle B'A'D'$, we have two congruent "small" spherical triangles by SSS. Therefore, we know that the angles are congruent as well, and thus, from #56, m($\angle A$) $\cong$ m($\angle A'$). Because from #56, we know that two congruent spherical triangles give us two congruent tetrahedra, we can conclude that $\triangle ABCD \cong \triangle A'B'C'D'$.

PROBLEM 58. Edges and Dihedral Angles

If $AB \cong A'B'$, $\angle AB \cong \angle A'B'$, $AC \cong A'C'$, $\angle AC \cong \angle A'C'$, $AD \cong A'D'$, $\angle AD \cong \angle A'D'$, then $\triangle ABCD \cong \triangle A'B'C'D'$.

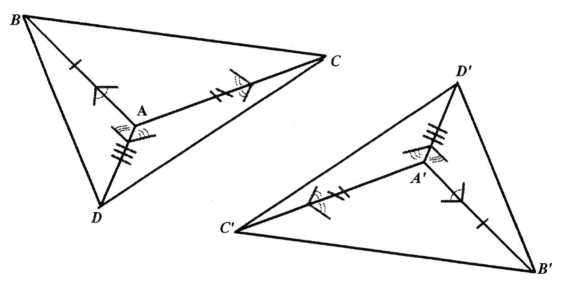

Outline of a proof:

Using AAA for the corresponding spherical triangles for angle A and A', we can conclude that the spherical triangles are congruent and so the solid angles, A and A', are congruent as well. Since the corresponding sides of the spherical triangles are congruent, the corresponding face angles with vertices A and A' are congruent, and we are then can proceed as in the proof of Problem 57.

Examples of Students' Work on Problem 58

Student answer:

Problem 58 - Amanda Cramer

I'd like to go backwards from what I did in Problem 57. If you know each of the dihedral angles are congruent, this means that the angles of the two spherical triangles are congruent. By AAA (for small triangles), the two spherical triangles are congruent, which means their sides are all congruent, which means that the two angles are congruent. From there you get all faces of the two tetrahedron congruent, so they're congruent.

PROBLEM 59. *Other Congruence Theorems for Tetrahedra*

Make up your own congruence theorems! Find and prove at least two other sets of conditions that will imply congruence for tetrahedra, that is, make up and prove other theorems like those in Problems 57 and 58.

We give here two such theorems. The students will come up with many more. See the Example of Students' Work for two other theorems.

FFF:

Two tetrahedra with three corresponding faces congruent are congruent.

By construction, three faces of a tetrahedron intersect at a vertex. If this is the case, we have enough information to determine the solid angle at that vertex. By Problem 57 or 58, the tetrahedra are congruent.

FAA:

Two tetrahedrons are congruent if they have one corresponding faces congruent as well as two solid angles, one of the angles not adjacent to the given face.

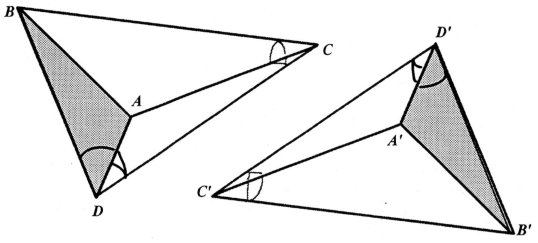

Using SAA, we can determine that $\triangle ABC \cong \triangle A'B'C'$ and $\triangle DBC \cong \triangle D'B'C'$. Three faces are then determined and, by FFF, the tetrahedra are congruent.

Locally, a 3-sphere feels like 3-space: the faces of a tetrahedron will be triangles in a 3-sphere. By what we know about the 3-sphere (Chapter 11) we know that a triangle in a 3-sphere always lies on a great 2-sphere. So we must consider tetrahedra which have as faces triangles that fit inside an open hemisphere of the great 2-sphere. In order to be able to use SAA the edges of the tetrahedra must be less than a 1/4 great circle. Thus:

FAA is true for tetrahedra in a 3-sphere if the edges are less than 1/4 of a great circle.

Examples of Students' Work on Problem 59

Student answer:

Problem 59 - Jason Fromberg

The first theorem is the *Base-Face-Angle Theorem*. We are given that:

$\angle BAC \cong \angle B'A'C'$, $\angle BAD \cong \angle B'A'D'$, $\angle DAC \cong \angle D'A'C'$
$\angle ABD \cong \angle A'B'D'$, $\angle ADC \cong \angle A'D'C'$, $\angle ACB \cong \angle A'C'B'$
$BC \cong B'C'$, $BD \cong B'D'$, $DC \cong D'C'$:

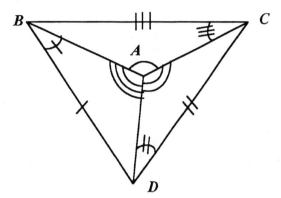

From Problem 57, we know that $\angle A \cong \angle A'$ because they have the same angles at vertex A. We also know that $\triangle BAD \cong \triangle B'A'D'$, $\triangle ADC \cong \triangle A'D'C'$, and $\triangle BAC \cong \triangle B'A'C'$ by the AAS Theorem. Therefore, since we know the included solid angle, we know $\triangle ABCD \cong \triangle A'B'C'D'$.

The second theorem is the *edge-edge-edge-edge-edge-edge theorem*, a.k.a. *(edge)^6 theorem*. Here we are given that all of the edges of the tetrahedron are congruent:

$BC \cong B'C', BD \cong B'D', DC \cong D'C', BA \cong B'A', AD \cong A'D', AC \cong A'C'$

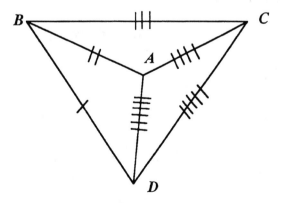

We know from SSS that $\triangle BAD \cong \triangle B'A'D'$, $\triangle ADC \cong \triangle A'D'C'$, and $\triangle BAC \cong \triangle B'A'C'$. We also know from SSS that $\angle BAC \cong$

$\angle B'A'C'$, $\angle BAD \cong \angle B'A'D'$, and $\angle DAC \cong \angle D'A'C'$. Therefore, from Problem 57 we know that $\angle A \cong \angle A'$. Since we have three faces and an included solid angle, we know that $\triangle ABCD \cong \triangle A'B'C'D'$.

PROBLEM 60. *The Five Regular Polyhedra*

Show that there are only five regular polyhedra. In Euclidean 3-space, to say "there are only five regular polyhedra" is to mean that any regular polyhedra is similar (same shape, but not necessarily the same size) to one of the five. It still makes sense on a 2-sphere to say that "there are only five regular polyhedra," but you need to make clear what you mean by this phrase.

Let us begin by following the hint given to the students. First, determine the angle of an *n*-sided regular polygon. An *n*-sided regular polygon can be inscribed in a circle, dividing the circumference of the circle into *n* equal arcs. If we connect the vertices of the polygon to the center of the circle, and if we connect by chords every two consecutive vertices of the polygon, we get *n* congruent isosceles triangles with one vertex at the center of the circle. The base angle of these triangles is equal to half the interior angle of the *n*-sided regular polygon. The angles with their vertex at the center of the circle in each of the congruent triangles have measure of $360°/n$. Thus, the sum of the base angles of the isosceles triangles is $180° - (360°/n)$, that is, $(n-2)180°/n$.

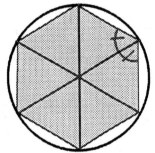 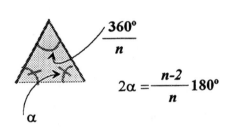

Now, the solid angle of a regular polyhedra is defined by the three adjacent sides which intersect at a vertex of the polyhedron. If the sum of face angles at that vertex is more than $360°$, then we do not have a solid angle, given that if the sum is more than $360°$, then the corresponding spherical triangle will not be contained in a hemisphere. Consequently, a polyhedron has to have faces which are regular *n*-gons with the number of faces less than the number of times $(n-2)/n$ fits into 360. That is, the polygonal face of a regular polyhedron has at most 5 edges. The faces of a regular polygon are then either a triangle, a square or a pentagon.

 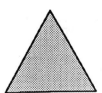 

Note that at each vertex the sum of the face angles has to be strictly less than 360°, otherwise the solid angle will be flat (corresponding to a hemisphere in a 2-sphere). Since a square's interior angle is 90°, we can join at most 3 square faces at a vertex. For a pentagon, the interior angle is

108°, and thus at each vertex the maximum number of pentagonal faces is 3. For an equilateral triangle the interior angle is 60° and thus we can have 3,4 or 5 triangular faces at a vertex. We have then 5 possible ways of arranging a regular polygons at the solid angle of a regular polyhedron. It remains to prove that each of these five ways produces precisely one regular polyhedron.

Since there are either 3, 4 or 5 faces at each vertex, the small spherical polygon which defines the solid angle at the vertex has 3, 4 or 5 sides and its angles are congruent to each other. In each case, the lengths of the sides of the spherical polygon are determined by the face angle of the polyhedron. In the three cases where the spherical polygon is 3-sided, that is a triangle, there is (by SSS) only one triangle possible. Thus the solid angles and dihedral angles are determined. In the two cases where the spherical polygon is 4- or 5-sided, look at the dual (see Chapter 16) of the polygon. Because the sides of the spherical polygon are determined, the angles of the dual are determined. Now subdivide by joining the center of the dual to each vertex. By AAA, the triangle in the subdivision are determined and thus the dual, the original spherical polygon and the solid angle are determined.

Thus, each of the five ways of arranging faces around a vertex of a regular polyhedron determines a unique face and solid angle. In Problem 57 we proved that a tetrahedron is determined if we have one of its solid angles and its faces. Now for the other polyhedra, we can follow the same steps that we did for the tetrahedron. That is, if we have two regular polyhedra with a congruent solid angle and congruent faces then the two polyhedra are congruent.

Examples of Students' Work on Problem 60

Professor and T.A. comments: *Student answer:*

Problem 60 - Jun Kawashima
If the faces are:

a.) regular quadrilaterals, each vertex angle is ninety degrees, so that there are at most three quadrilaterals (<360 / 90) intersecting at each vertex to create a solid angle. If four quadrilaterals intersected, then the result will be a plane. On the other hand, if there were only two faces, then there wouldn't be an enclosure. Thus, there must be precisely three faces at each enclosure.

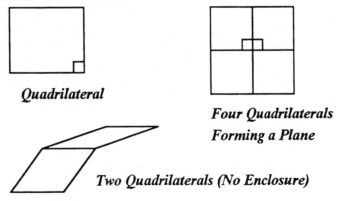

Quadrilateral

*Four Quadrilaterals
Forming a Plane*

Two Quadrilaterals (No Enclosure)

b.) regular pentagons
Maximum number of faces at each vertex < 360 / ((180·3) / 5) = 10/3.

So, there are at most three faces at each vertex. Using the same argument as before, there would be no enclosure if only two faces intersected at each vertex. Again, three faces intersect at each vertex.

c.) regular hexagons

Maximum number of faces at each vertex $< 360 / ((180 \cdot 4) / 5) = 2.5$.

Having two faces was impossible, so we cannot use hexagons.

d.) equilateral triangles

Maximum number of faces at each vertex $< 360 / 60 = 6$.

$2 <$ number of faces $< 6 \Rightarrow$ have 3, 4, or 5 intersections.

Thus, any other regular polygons having more than five sides cannot form the solid angle of a regular polyhedra, so that we are left with the possibilities:

equilateral triangles - 3 cases - tetrahedron, octahedron, and icosahedron.

regular quadrilaterals - square [cube].

regular pentagon - dodecahedron.

Why is it that the regular polyhedron's shape is determined by one solid angle and a face?

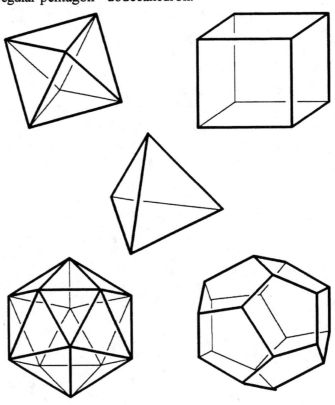